LOST INDUSTRIES *of* SARATOGA COUNTY

LOST INDUSTRIES *of* SARATOGA COUNTY

Timothy Starr

The History Press

Published by The History Press
Charleston, SC 29403
www.historypress.net

Front cover: Town of Clifton Park History Collection, Clifton Park–Halfmoon Public Lbrary. *Back cover*: *top left*: Town of Halfmoon Historian's Collection; *top right*: Stillwater Town Historian; *bottom*: Courtesy of the Corinth Town Historian.

First published 2010

Manufactured in the United States

ISBN 978.1.60949.079.9

Library of Congress Cataloging-in-Publication Data

Starr, Timothy, 1971-
Lost industries of Saratoga County / Timothy Starr.
p. cm.
Includes bibliographical references and index.
ISBN 978-1-60949-079-9
1. Manufacturing industries--New York (State)--Saratoga County--History. 2. Saratoga County (N.Y.)--History. I. Title.
HD9727.S27S73 2010
338.09747'4809034--dc22
2010035545

Contents

Acknowledgements

The author would like to thank the following individuals and organizations for their valuable contributions: the New York State Archives; the Ballston Spa Public Library; the Mechanicville Public Library; the Saratoga Springs Public Library (Saratoga Room); Joy Houle at the Brookside Museum, Saratoga County Historical Society; Brad Utter at the Waterford Historical Museum and Cultural Center; Rachel Clothier at the Corinth Museum; Gail Winters at the Clifton Park–Halfmoon Public Library; Linda Palmieri and Linda Sanders, Stillwater town historians; Paul Loatman, Mechanicville historian; the Saratoga County historian; Priscilla Edwards, Edinburgh historian; Kristina Saddlemire, Schuylerville historian; Chris Morley, Ballston Spa historian; Paul and Marilyn Pastore; and Pat Streifert.

Introduction

A visitor to Saratoga County would likely become confused by the presence of huge, vacant buildings scattered across many of its villages and hamlets. Why were these structures built? What purpose did they serve? Why were they abandoned? The answers to these questions lie in studying the industrial history of Saratoga County, a fascinating era that transformed the local economy and affected the lives of many hundreds of thousands of people around the world.

The presence of rare mineral springs in Saratoga and Ballston Spa drew the first permanent settlers in the late 1700s. A growing need to transport goods to New England and the western frontier prompted the state to build the Champlain and Erie Canals, both of which bordered Saratoga County. The burgeoning tourism trade of Saratoga Springs later prompted investors to build railroads through the county. The presence of these two transportation systems, abundant forests and several powerful rivers led to the establishment of a variety of manufactories throughout the nineteenth century.

It is safe to say that without the Hudson, Sacandaga and Kayaderosseras Rivers, very little industrial development would have occurred. All of the county's manufacturing centers were built on the banks of these waterways in order to take advantage of the power that could be generated for machinery. Such communities as Corinth, Schuylerville, Mechanicville, Waterford, South Glens Falls and Stillwater owe their very existence to their rivers. Towns that did not have significant rivers—such as Clifton Park, Charlton, Malta and Galway—never developed significant manufacturing economies

Saratoga County, circa 1900. Today the county consists of nineteen towns, nine incorporated villages and two cities (Saratoga Springs and Mechanicville). *Courtesy of the author.*

and only experienced increases in population during the twentieth century when automobile transportation became feasible.

The Capital District of New York State, comprising Albany, Rensselaer, Schenectady and Saratoga Counties, was a strategic and important region for more than a century after the Revolutionary War. Albany was near the end of the navigable portion of the Hudson River and, therefore, acted as a staging area for goods from the north and west heading to the international port of New York City. The Champlain Canal linked the Capital District with points north to Canada, while the Erie Canal followed the Mohawk River and linked the Capital with Buffalo and the West.

Albany and Troy developed a thriving foundry industry beginning in the early 1800s that included the manufacture of iron, steel and stoves. This was made possible by iron deposits in surrounding counties, molding sand along the Mohawk Valley, ample fuel supplies from the Adirondacks and a direct water route to New York City. By the end of the century, the steel foundries and stove manufacturers in the region were known around the world.

Saratoga County industry could not compete with the foundries located near its southern borders, so other industries were developed that took advantage of local natural resources. The most successful of these were the paper mills. The presence of millions of acres of woodland close at hand provided nearly unlimited fuel supplies for early paper mills and other factories. When the wood pulp and sulphite processes were developed that allowed the mills to use wood for paper instead of cotton rags, the paper manufacturers expanded even further. Woodenware factories, planing mills and tanneries also benefited directly from inexpensive timber that was hauled in from just a few miles away.

The first "industry" in the county was that of subsistence farming, as early settlers were preoccupied with raising enough food to feed their families. The increased use of mechanical technology allowed farmers to grow cash crops such as wheat, hay and oats. The introduction of clover and alfalfa, which delivered nitrogen from the atmosphere into the soil, allowed the restoration of fields that had been in continuous cultivation for decades. As transporting goods became easier and faster, large farms from the West drove many local farms out of business.

One of the county's earliest gristmills was located in the town of Malta. Although the millstones were found many years later, the only proof that it existed before 1800 are stories passed down through the generations. Michael and Nicholas McDonald were among the first permanent settlers in the area, arriving in 1763. It is related that the brothers brought their grain across Ballston Lake in a canoe and carried it to the Malta mill to be ground into flour. After farming, flour milling was recorded as the leading manufacturing industry before the Civil War. Sawmilling, by comparison, was valued in sixth place.

All early mills were powered by small dams, such as this one on the Milton–Saratoga Springs town line. *Courtesy of the author.*

The millstone was the most valuable component of early gristmills. The Palmer family of Stillwater purchased a stone from France, said to be one of the best in the area, and put it to good use in their mill during the 1770s. When war broke out with the British, the Palmers feared that the stone would fall into enemy hands. Since the wheel was so valuable, they had to come up with an alternative to destroying it. One night, as General Burgoyne advanced toward Stillwater, Elias Palmer and several men loaded the stone onto a specially constructed raft. After managing to get two-thirds of the way across the Hudson River, they found a hole ten feet deep and dumped the wheel into it. When the war was over, the wheel was raised from its hiding place and put back into the mill, where it served for many years. Some believe that when the gristmill was torn down, the stone was placed near the road and buried under the fill, where it lies to this day.

The first known sawmill in the county is thought to have been constructed at Halfmoon in 1762, followed by mills in Northumberland (1777), Charlton (1783), Wilton (1784), Providence (1786), Greenfield (1789) and Hadley (1791). According to the census of 1860, there were about twenty-

Daniel Dow's sawmill in Clifton Park Village was typical of those that served the county. *Courtesy of the George Dows Collection, Town of Clifton Park, Clifton Park–Halfmoon Public Library.*

two sawmills in the county that mostly served local residents, but by 1870, there were over forty, many serving the new paper mills and woodenware factories. Timber from farms was sold to others for firewood, charcoal, ashes, building materials and even tool handles. Full-time sawmills turned out dressed lumber for sale to wagon makers, coopers, boat and bridge builders, woodenware manufacturers and coffin makers.

Although the manufacturing sector of Saratoga County never exceeded those of Albany and Rensselaer Counties, it employed many thousands of people and circulated millions of dollars into local economies. Besides those that depended on a ready supply of timber, a wide variety of industries was established, including the manufacture of fire engines, clothing, glass and medicines.

The sheer magnitude of the documented manufacturers makes it impossible to describe them all in any detail. There are also many dozens of manufacturing operations that were short-lived and never documented. Most of the early sawmills and gristmills were built by farmers with a bit of extra cash, served a few local families and then disappeared without a trace.

Small tanneries were often set up in the woods and dismantled once the trees were gone. Many of the earliest mills in Halfmoon and Clifton Park, which were settled before the American Revolution, are only known through traditions and oral histories before being set down on paper much later.

Published histories of Saratoga County nearly always focus on the famous Revolutionary War battles, mineral springs, horse racing and the gambling era. When authors write about industrial history in the Capital Region, the focus is on Troy and its iron and collar industries and Schenectady, which hosted the American Locomotive Company and the headquarters of General Electric. The industries of Saratoga County seem to have been largely forgotten by the general public since the decline of manufacturing in the early 1900s. Only by studying records from dozens of libraries and town historian offices could its lost history be "rediscovered." It is hoped that this compilation will help new generations of residents understand what life was like 150 years ago and the importance of Saratoga County to the rest of the nation during its industrial era.

THE IMPORTANCE OF WATER POWER

For decades after Saratoga County was first settled, the advent of electrical power was far in the future, and steam was just becoming a viable alternative. Therefore, the existence of a river to power industry was a necessity for pioneer businessmen. All early industrial pursuits were powered by water wheels that were located on either river rapids or dams. The existence of several suitable rivers for industry in Saratoga County meant that larger settlements could be supported and business centers developed.

Almost without exception, industry before 1800 was established along the Hudson, Sacandaga or Kayaderosseras Rivers or their tributaries. The Hudson River provided for the settlement of such villages as Mechanicville, Stillwater, South Glens Falls, Waterford and Corinth. The Kayaderosseras provided a multitude of suitable sites in the town of Milton where such villages and hamlets as Ballston Spa, Milton Center, Factory Village and Rock City Falls became early industrial centers. The Sacandaga River gave life to Batchellerville and Beecher's Hollow in the town of Edinburgh and Conklingville in the town of Hadley.

A comprehensive study of the nation's water powers by Professor W.P. Trowbridge in 1885 gives historians an accurate firsthand account of how industries utilized this natural resource during the peak years of manufacturing in Saratoga County, and it will be referred to below. Logging operations had been the first and most important use of the river beginning before the Revolutionary War. Untold millions of logs were floated down the Hudson and several of its tributaries to sawmills in South Glens Falls. Few dams of any size were constructed along the Hudson until the mid-

1800s, and no effort was made to coordinate these few dams to hold back flooding in the spring or to release water during the summer months. In fact, dams north of South Glens Falls would have been a nuisance for sawmill owners who depended upon a clear path for floating logs down from the Adirondack wilderness.

However, the economic expansion that occurred after the Civil War prompted a flurry of activity along the Hudson as entrepreneurs sought to exploit the river for power generation. One of the most ambitious dam projects of that time was prompted by the failure of the Glen Paper Collar Company in Ballston Spa. One of that enterprise's owners, a young inventor named Horace Medbery, sought new business opportunities and came upon an undeveloped section of the Hudson in Mechanicville. He and several other partners created the Hudson River Water Power and Pulp Company for the purpose of erecting a giant dam and what they hoped would be some of the largest pulp and paper mills in Saratoga County nearby. Work on the dam commenced in 1882. When it was completed, the dam measured 1,000 feet long, 16 feet high, had a gateway 115 feet wide and contained 9,000

The large dam at Mechanicville, built by the Hudson River Water Power and Pulp Company, generated horsepower for several mills. *Courtesy of the New York State Archives.*

yards of stone. A tramway was set up extending the length of the structure to deliver materials. When the dam was completed, it increased the potential horsepower from 182 at average river flow to over 3,600 with the 20-foot fall.

About a mile above Mechanicville was Howland's Dam, utilized by the Saratoga Paper Company for the manufacture of straw paper, although at the time of the 1885 study it was in a "very dilapidated" appearance, having sunk in several places to give it an uneven and irregular look. The next significant use of the river to the north was at Stillwater where the Hoosac River empties into the Hudson. The dam at that point was fifty years old but had been kept in good working order by the multitude of mills that utilized it. The dam contained thirteen gates, each having two upright posts with iron racks, and had a bulkhead constructed with timber. A short side canal fifteen feet wide opened from the pond above the bulkhead and connected with a larger canal one hundred feet wide that followed the river for one-third of a mile. All of the mills that utilized the dam were built over the canal supported by stone piers. These mills were listed as follows:

John B. Newland gristmill, sixty horsepower
Newland knitting mill (hosiery), eighty hands, seventy horsepower
Mosher, Judd and Company paper mill, fifteen hands, one hundred horsepower
D. and W. Pemble's straw-board paper mill, six hands, forty horsepower
R. and H. Newland's knitting mill (hosiery), sixty hands, forty horsepower
Denison and Co.'s knitting mill (hosiery), sixty hands, forty horsepower

A total of 350 horsepower was thus being used, and it was reported that no shutdowns had ever been necessary due to low water. It was noted that on one occasion the mills were stopped by backwater from a heavy ice jam, but this was rare. Dam owner John Newland sold power to the mills by the inch, an inch consisting of how much water could pass through an opening one inch square under the head of water at each mill.

Next was the Saratoga Dam where the Champlain Canal crossed the Hudson and where the canal was fed enough water to carry it to Waterford. This dam measured eight feet high and was completed by New York State in 1873. The sawmill that utilized the power at this dam and the improved privileges upstream from Sandy Hill and Fort Edward were located in Warren County.

At South Glens Falls was located the extensive sawmills and limekilns of the Morgan Lumber Company, and although it was estimated that 500

Nearly all the industry of South Glens Falls was concentrated on the Hudson River. On the left is the Glens Falls Paper Company. Warren County is on the right. *Courtesy of the New York State Archives.*

horsepower could have been generated at the fall there, it had yet to be developed. Morgan Lumber was the primary user at this point, sawing as much as thirty-five million feet of lumber per year, but only required a small crib dam to power its mill. The primary purpose of the dam was for use as a sluiceway to process logs floating in from the north and separating out those that were destined for Sandy Hill and Fort Edward. The only other mills on the south bank of the river (and thus located in Saratoga County) to use the water power at this point were the Lapham and Company gristmill, requiring 250 horsepower, and the Glens Falls Paper Mill Company.

One mile north of South Glens Falls was the state Feeder Dam, constructed of stone held by crib-work. A double-lock provided a side for boats and another side as a waterway used for running over logs. Its name was derived from the fact that it "fed" the Champlain Canal a large amount of water from the Hudson in order for it to operate properly. Located at this site were the sawmills of Finch, Pruyn and Company; Van Deusen and Freeman; and Rugg, Sherman and Company, where some 350 men were employed. Just above the Feeder Dam was a long bend in the river where the great "Hudson River Boom" of thousands of logs was held for the use of the mills each year.

The last Hudson River privilege within the borders of Saratoga County was located near the town of Corinth at Palmer Falls. At that time, the power was owned by a group called the Palmer Water Power Company, which developed that section in an effort to attract entrepreneurs to Corinth. It was one of the most promising parts of the river for power generation, with high falls and a bed of solid rock. Although it was some years before any manufactory took the bait, the Pagenstecher brothers finally arrived and built their huge paper mill (called the Hudson River Pulp and Paper Company) a short distance away from the Palmer Water Power Company. Although the dam was rebuilt later, in 1885 it consisted of a raceway six hundred feet long and from ten to thirty feet high, abutting a rocky wall cut out by the river. Water was admitted to the race through a timber bulkhead measuring fifty feet across and containing four gates. The dam, race and connected work were all constructed at a cost of $35,000. Prior to the Hudson River Pulp and Paper mill, only a few small industries, such as sawmills, gristmills and a woolen mill, had used the water power in Corinth, even though the effective potential horsepower was the third largest of any other location (as noted above, the dam at Mechanicville had the potential of 3,600 horsepower, and at Glens Falls it exceeded 2,000 horsepower, compared to 1,500 at Corinth).

The Hudson was far from being the only water source in the county. Its greatest tributary, the Mohawk, provided Waterford with a means to create the King's Canal, constructed by John F. King in the 1830s. Water was admitted into the canal through a timber bulkhead almost fifty feet long and passed a half mile down the north bank of the river to supply various mills with 450 horsepower. Some of the manufacturing along the canal included fire engines, stocks and dies, strawboard, flour and machine work.

Another Hudson River tributary of importance to the county was Fish Creek, fed by Saratoga Lake, which is in turn fed by the Kayaderosseras Creek. The first privilege on Fish Creek was owned by the storied Bullard family at Schuylerville, operators of an important paper mill. The 9-foot-high dam was constructed of stone and cement and measured 140 feet long. The raceway was open for 25 feet before passing underground to the mills. Other businesses in that section were the small foundry of Craw and Dennis, a five-run gristmill and a sash and blind shop. The Bullards also owned a small dam on the nearby Champlain Canal for operating a sawmill that was once owned by the village's namesake, General Phillip Schuyler.

A short distance above Schuylerville was the extensive dam and operations of the Saratoga Victory Manufacturing Company cotton mill. It consisted of two mills—the Horicon Mill at Schuylerville, with over ten thousand

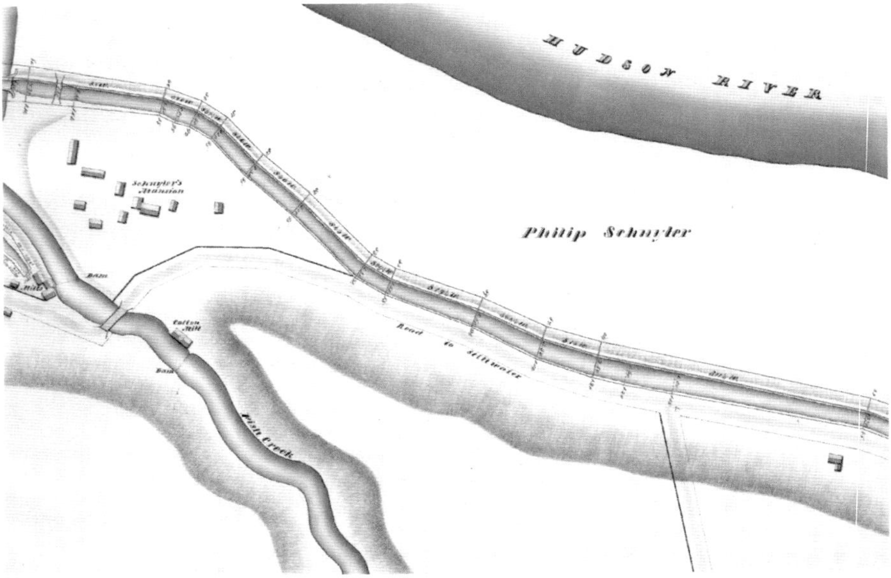

A map of Schuylerville dated 1830 shows the mansion and mills of Phillip Schuyler II a few years after the Champlain Canal was built. *Courtesy of the New York State Archives.*

spindles, and the Victory Mill about one mile above, with twenty-nine thousand spindles. Because there was usually a shortage of water for three to four months of the year, steam had to be utilized for auxiliary power. The Horicon Mill dam was a timber structure 19 feet high that powered a 160-horsepower wheel. The Victory Mill dam was also made of timber, measuring 140 feet long and 8 feet high. Water was conveyed into a 500-foot-long race that powered an 800-horsepower wheel. The wheel pit was sunk 30 feet into solid rock, and the tail race (where the spent water is emptied back into the river) extended 170 feet through a tunnel 13 feet wide and 7 feet high and another 500 feet of open channel. The only other dam on the Fish Creek was at Grangerville that powered a saw- and gristmill.

The other two rivers of importance were the Sacandaga and the Kayaderosseras. The former originates primarily from several lakes in Warren and Fulton Counties and flows through Saratoga County in the towns of Edinburgh, Day and Hadley. About one mile west of where the Sacandaga empties into the Hudson was the first dam of importance, owned by the Sacandaga Pulp Company and measuring 275 feet long. It was described as being "very leaky" and constructed from logs resting on a sandy base. Nearby was another dam utilized by the New York Pulp

Company, built with logs that were filled in with stone. The first dam was destroyed by a flood that also carried away the mill situated on top of it; the replacement dam was of sturdier construction, and the mill was situated farther up the shore.

The next dam to the west was located in Conklingville and thought to be built about 1830. It was 240 feet long and contained crib-work abutments with smooth timber facing. Clement's woodenware factory was on the left bank, while a raceway several hundred feet long was on the right bank and led to Henry Poor and Son's tannery. Farther west in Edinburgh were located several woodenware factories and other small industries, but no significant dams were mentioned beyond Conklingville.

No industry of note operated on the Kayaderosseras Creek between Ballston Spa and Saratoga Lake, but in Ballston Spa and north to the Greenfield town line was a wide variety of manufacturing concerns that utilized the creek for power. In Ballston Spa proper were several tanneries, the Blue Mill and several paper mills owned by George West. North along the creek was a succession of manufactories including a foundry, sawmills, the hard edge tool factories of Isaiah Blood and about a dozen paper mills. In the short span of fewer than ten miles, the creek resembled a series of lakes as dams were utilized in the hamlets of Bloodville, Factory Village, Milton Center, West Milton, Rock City Falls and Middle Grove. The creek was easily the single greatest factor in the transformation of Milton into what was then considered to be a heavily industrialized town.

Canals and Railroads

No single factor was more important to the development of heavy industry in the county than the construction of canals and railroads in the early part of the 1800s. Industrial activity until the 1830s, with a few exceptions, was limited to serving local populations with processed food, clothing and lumber. The arrival of these new transportation networks allowed entrepreneurs to ship goods from Saratoga County to Albany and thence around the country more cheaply and reliably than ever before.

New York governor George Clinton first proposed that canals be built between the Hudson River and Lake Ontario and between the Hudson and Lake Champlain. After much discussion, legislation and wasted efforts, a commission was created in 1819 to begin a formal survey for the Erie Canal. Construction finally began and was completed on October 26, 1825. The entire canal was 363 miles long, forty feet wide and four feet deep. In Saratoga County, the canal crossed the Mohawk River from Schenectady via an aqueduct in the town of Clifton Park and followed the banks of the Mohawk on the southern end of both Clifton Park and Halfmoon. It then recrossed the Mohawk at Crescent and entered Albany County. The canal was widened in 1839 to handle larger boats, by which time there were two aqueducts, four locks, a feeder canal and nearly twenty bridges along the 12-mile stretch in Saratoga County. Additional modifications were made at the turn of the century when a barge canal was constructed from Waterford to Rexford.

Although the Erie did not extend very far into Saratoga County, it was of great importance to Halfmoon and Clifton Park, which until that time had been limited to agricultural pursuits. It was an important shipping lane for western states to ship their products to New York City, and all of it passed through Saratoga County. In addition to the increased activity along the southern border of the county, several substantial communities were established, such as Ushers and Vischer Ferry. Boat building and sand mining were the chief industries, but several mills were erected to support the growing population base.

The sleepy little hamlet of Crescent was dramatically altered by the arrival of the Erie Canal. Consisting of only a few small houses prior to 1820, the hamlet quickly became a hub for shipping grain, ice, bricks and molding sand to the Capital District and New York City. Stores and industry sprouted up as the population grew, until Crescent warranted its own post office, school, hotel and newspaper. A twenty-six-arch aqueduct was constructed nearby to carry the canal from one side of the Mohawk River to the other.

The Champlain Canal was of greater importance to the development of the county, as a greater portion of it lay within its borders. When completed in 1823, the canal followed the Hudson River and passed through the towns of Waterford, Halfmoon, Stillwater and Saratoga, crossing into Washington County near the boundary with Northumberland. Thus,

Clute's Dry Dock was established on the Erie Canal in Vischer's Ferry to build and repair canalboats. *Courtesy of Town of Clifton Park History Collection, Clifton Park–Halfmoon Public Library.*

View of the Champlain Canal looking east toward the Hudson River in 1898. *Courtesy of the New York State Archives.*

the villages of Waterford, Stillwater, Mechanicville and Schuylerville all derived great benefit from its shipping capacity. The cost of building the canal was relatively modest—only $875,000—compared to $7 million for the Erie Canal. A feeder dam was built in Glens Falls to "feed" the canal with water from the Hudson River. This gave the canal so much water flow that the Hudson River itself was not used for navigation thereafter until it reached Albany.

Schuylerville in particular saw a dramatic change once the canal was completed. Canal basins were constructed at various points along the canal so that boats could be turned around, loaded and unloaded. One of these basins was located in Schuylerville thanks to the efforts of Phillip Schuyler II, who deeded a right of way to the canal from his extensive landholdings. Tons of potatoes grown by local farmers were stored in Schuylerville warehouses until the fall, when they were loaded on dozens of canalboats. The new activity prompted merchants, mechanics, laborers and innkeepers to set up shop in such numbers that the once sleepy little hamlet was incorporated as a village in 1831. Parts of the canal were abandoned in the early 1900s when the Barge Canal was constructed, which combined the larger parts of the Champlain Canal and Hudson River.

Mechanicville was one of the largest terminals on the new Barge Canal, with a five-hundred-foot concrete dock for loading and unloading products. Thousands of tons of sand from the surrounding neighborhood were loaded onto barges. Lumber from the Pruyn Lumber Company was shipped around the world from this dock, while thousands of cords of pulp wood were unloaded for use at the West Virginia Pulp and Paper Company mill.

The village of Waterford became a major hub for shipping once the two canals were completed, since it was located at the southern terminus of the Champlain Canal and near the end of the Erie Canal. New buildings were constructed along the banks of the canals and surrounding streets consisting of canal worker residences, stores, warehouses, stables, boardinghouses and saloons. The area near Fifth and Broad Streets became known as "Canal Town." Several manufacturing plants were established along the banks as well, such as the Button Fire Engine Works and the Franklin Ink Works.

In 1828, a third canal was added to Waterford's extensive system. A canal expert named John Fuller King oversaw the construction of the King's Power Canal along the Mohawk River. With a drop of eighteen feet in just half a mile, the canal was ideally suited to generate water power for industry. It was considered one of the greatest engineering projects of its time and was responsible for the development of the Northside and Dial City sections of the village. A wide variety of mills was built along its banks, including textile mills, foundries, flour mills and distilleries. Some of the factory owners built their residences along Saratoga Avenue, which became known as "Mill Owners' Row."

Prior to the completion of the canals in Waterford, a small, two-hundred-yard pier served as an unloading point for vessels coming up the Hudson River. New facilities were needed once the canals were opened. At first, these consisted of wooden docking walls, but these were gradually upgraded with concrete walls and floating docks that became known as Waterford Harbor. Barges were then able to deliver huge supplies of iron ore and timber from the Adirondacks and marble from Vermont, among other commodities.

Canals were not the perfect solution. Although they dramatically lowered the cost of sending goods to the west and south, they were still exceedingly slow. A canal trip from Albany to Schenectady often took more than twenty-four hours, as boats had to make their way through an extensive lock system. Faced with this timetable, most passengers decided to take the two-hour stagecoach, even if it was a much bumpier ride. Clearly, there was a need for a faster and more efficient mode of transportation. Fortunately, the people of Saratoga County and the rest of New York State did not have long to wait.

The lasting influence of the canals that bordered the county was largely curtailed by the arrival of the railroads. The first railroad in New York State was built between Albany and Schenectady. Although it was not as financially successful as its investors hoped, it opened the eyes of many who realized the vast potential of railroads to transport people and goods. In 1831, the state legislature granted a charter to the Saratoga and Schenectady Railroad Company to construct a railroad between Schenectady and Saratoga, passing through the industrial village of Ballston Spa.

A short time later, businessmen in the city of Troy incited legislatures to approve the construction of a railroad of their own to Ballston Spa to keep Albany from monopolizing the northern trade. By October 1835, the first train on the Rensselaer and Saratoga Railroad had departed First Street in Troy for its trip to Front Street in Ballston Spa. Despite its name, the railroad only reached as far north as Ballston Spa, but soon after its completion an agreement was made with the Saratoga and Schenectady to use its trackage to Saratoga.

Throughout the late 1800s, a series of new railroads was built, such as the Saratoga and Washington, the Fitchburg and the Adirondack. Many of these became consolidated later in the century into the Delaware and Hudson Railroad system, which eventually spanned the eastern part of New York State and beyond (the Fitchburg later became the Boston and Maine). Other railroads, such as the Mount McGregor and the Saratoga Lake Railway, served only the tourist trade in Saratoga Springs.

Several communities around the county benefited directly from the railroads, especially those that were a stop on one of the lines. The extensive railroad facilities in Mechanicville employed six hundred men. Each day, forty-seven passenger trains arrived and departed over the Delaware and Hudson line, while eleven more departed over the Boston and Maine Railroad. The length of the Delaware and Hudson sorting yard alone measured nearly three miles exclusive of private sidings and had the capacity to hold 1,500 freight cars.

By the turn of the century, both steam and electric railroads crisscrossed the county. The Delaware and Hudson passed through Waterford and Mechanicville before heading east toward Ballston Spa and then north to Saratoga. The Adirondack branch spanned from Saratoga to the northern settlements of Greenfield Center, Kings Station, South Corinth, Corinth and Hadley before entering Warren County. The mainline left Saratoga and headed northeasterly to Gansevoort, Moreau Junction and South Glens Falls. The old Saratoga and Schenectady line still operated from Schenectady through Ballston Lake and Ballston Spa. The Boston and Maine system entered the county from the east at Mechanicville and turned north to Schuyler Junction (located just north of Saratoga Lake), where it split in two, going west to Saratoga and east to Schuylerville. After heading north to Northumberland, it entered Washington County. Another B&M line headed west from Mechanicville and passed through Ushers and Elnora before entering Schenectady County. The Hudson Valley Railway and the Schenectady Railway electric trolley lines often followed the steam railroads, serving the same population centers.

The railroads did not automatically lead to industrial development. Hamlets such as Kings Station, Moreau Junction, Ballston Lake and Elnora never developed heavy industry, even if their residents were able to travel about much easier and farmers were able to export their crops much less expensively than before. Rather, the railroads enabled those areas where water power was present to establish new mills or enhance those that already existed.

Extensive railroad yards and facilities were constructed throughout Saratoga County, such as Saratoga Springs (pictured) and Mechanicville. *Courtesy of the Milton Town Historian Collection.*

The railroads were perhaps the deciding factor in the large-scale growth of the paper industry in the county. It would have been impossible for the large mills of Corinth, Ballston Spa, South Glens Falls, Schuylerville and Mechanicville to compete with others in the country if there was not an inexpensive and reliable way to feed them raw materials and deliver finished goods to New York City or the West. Railroads were deemed so important that a unique electric railroad was built in 1896 for the sole purpose of serving the paper mills and other industries along the Kayaderosseras Creek in Milton. The Ballston Terminal Railroad interchanged with the Delaware and Hudson near Blue Mill Pond and followed the creek northward, serving mills in Bloodville, Factory Village, West Milton, Rock City Falls and Middle Grove.

Based on the overall success of certain villages within the county, it is apparent that railroads were second in importance only to the existence of water power for industry. Some settlements, such as Gansevoort, had a railroad connection but little water power. Others, such as Beecher's Hollow and Batchellerville, had ample water power but no railroad connection. Those with one but not the other never developed large-scale, long-lasting industry. It is no coincidence that Waterford, Corinth, Ballston Spa, Schuylerville, South Glens Falls and Mechanicville all had significant water powers and direct rail links to the outside world, and all possessed industrial-based economies for over a century.

Town and Country

It is almost astonishing how many towns in Saratoga County developed an industrial base that exported goods outside of the community. For example, in the more heavily populated county of Albany, nearly all exporting industry was centered in the cities of Watervliet, Cohoes and Albany. In Saratoga, beyond the obvious industrial economies of Waterford, Mechanicville, Corinth and Ballston Spa, several other towns developed unique and sometimes significant manufacturing enterprises that have been all but forgotten. These included the woodenware factories of Edinburgh, the sand harvesting operations of Clifton Park, the sawmills of Moreau and the mineral water bottling operations of Saratoga Springs.

Located at the merging of the Mohawk and Hudson Rivers, Waterford was always ideally suited for the placement of water-powered industries. Spafford's 1813 *Gazetteer of New York* claimed that the town was the most populous in the county with 190 houses and stores. Although Ballston Spa experienced significant commercial development after the railroad arrived, Waterford appears to have had a greater variety of manufacturers. Some early industries included small flouring mills, sawmills, a tannery and blacksmith shops. Around 1825, a rather extensive cooperage business was established that turned out thousands of tubs and barrels per year, partly to support Albany's brewery industry. The construction of the King's Canal induced several manufacturing concerns to relocate from nearby cities, such as a cotton factory, a twine factory, the Kilby and Vanderwerker foundry, the Colonel Olney Machine Shop and the Eddy ink works.

A complete listing and history of the manufactories of Waterford would itself make a sizeable book. The textile industry in Waterford was by far the most extensive in the county and even rivaled nearby Cohoes and Troy in some years. Former Waterford town historian Dennis Rivage compiled a list of over fifty knitting mills that were started up between 1870 and 1920, some of which employed over five hundred operatives. Other businesses included the Button Fire Engine Company, the Brooks Nut Factory, the Waterford Soap and Candle Factory, the Rock Island Flouring Mill and the Van Kleeck brush-making factory. Foundries were important from an early date and included the long-lived Gage Machine Works, Holroyd and Company, the J.M. King foundry and the Globe Iron Works. The Mohawk and Hudson foundry of George Eddy was particularly successful thanks to his invention of water valves and the world-famous Mohawk Fire Hydrant.

The city of Mechanicville is a relatively modern settlement when compared to others in Saratoga County. Despite the presence of the famous King's Highway between Albany and Canada through the settlement, the

Carriage shops like this one on old Route 146 in Clifton Park were scattered throughout the county. *Courtesy of the Town of Clifton Park History Collection, Clifton Park–Halfmoon Public Library.*

This structure in Vischer's Ferry was built by William Peters in 1814 and used as a sawmill, gristmill and woolen mill. *Courtesy of the Howard Becker Collection, Town of Clifton Park, Clifton Park–Halfmoon Public Library.*

population and industrial activity remained small until the arrival of the Champlain Canal and the Rensselaer and Saratoga Railroad. These two transportation routes provided the means for Mechanicville to establish major industry and support a larger population. The first industry of any importance was the manufacture of thread, established by Samuel Chase and Lewis Smith. The American Linen Thread Company derived its power from the Anthony Kill and proved to be a success for many years, employing up to 175 people.

For several decades, the county's paper industry was by far the largest, both in terms of production and number of people employed. A group of investors organized the Hudson River Water Power and Pulp Company in 1882 at the northern end of Mechanicville. Their goals were to build a large dam over the Hudson River, erect a fiber and paper mill and sell the excess power generated by the dam to other parties. Although the original partners never accomplished all of these goals, the dam was built and a paper mill was situated nearby. Later the mill was purchased by the West Virginia Pulp and Paper Company, which greatly expanded its capacity.

One of the largest sash and blind mills was also located in Mechanicville. Jacob LaDow owned a sawmill in Stillwater under the name Cornell and LaDow, where Albert Barnes became an apprentice at the age of fifteen. Barnes eventually moved to Mechanicville and formed a partnership with Jacob's son Daniel. Prior to his interest in the sash and blind business, Daniel LaDow spent time in Waterford with the firm Allen, Stewart and Company in the knit goods business. He then came back to Stillwater and established a knitting mill of his own that he operated until 1868, by which time he was ready to join his father's sash and blind business. The following year, he took over his father's share and began a successful partnership with Albert Barnes. In 1885, the factory burned to the ground, but it was rebuilt on a larger scale near the railroad station. Building after building was added until the business became one of the most complete of its kind in the county, earning $200,000 per year. The main factory was a two-story brick structure measuring one hundred by twenty feet that contained a sixty-horsepower engine. The large warehouses and mill yards were conveniently located between the canal and the railroad. Up to seventy men were employed manufacturing doors, sashes, moldings, lathe, shingles and King's Windsor cement.

As with Waterford and Ballston Spa, the production of clothing was the last wave of industry in Mechanicville, and one of the most profitable. The first such enterprise was a shirt factory established by Charles Crosby on First Street. This later became the E.E. Ide and Company Shirt and Collar Factory. The plaster mill of the American Linen Thread Company came under the ownership of the Knickerbocker Hosiery Mills, owned by William Neilson, Rial Newland and Henry Newland. About one hundred men and women were employed manufacturing men's and women's all-wool underwear at a rate of over 350,000 per year. Another shirt factory was operated by Hall and Hartwell on Mabbett and Canal Streets. Kay Dunhill, Inc. commenced the manufacture of a line of dresses in 1937, employing eighty people. Kay Dunhill would be the last textile firm in the city, closing down in the 1980s.

The water power furnished by the Hudson River in the town of Corinth had great potential for early explorers to utilize, but extensive development did not begin until well into the nineteenth century. Niagara Falls native Thomas Brown purchased property near the Hudson in 1859 and constructed an edge tool factory with a canal race that fed water from the river to his shop. After only operating the factory for a few years, a lack of iron supplies stemming from the Civil War forced him to convert it to a woolen mill. The property was later sold to the Hudson River Pulp and Paper Company. Other industry was developed south of the river, especially

The Elixman Paper Core Company of Corinth manufactured large paper rollers for International Paper until 1976. *Courtesy of the Corinth Town Historian.*

in a settlement called Chapmanville, later known as South Corinth. The Parkman gristmill was in operation by 1800, followed by the woolen mill of Washington Chapman a few years later. The mill was purchased by Morgan Prentiss mid-century for the manufacture of bolts, nuts and charcoal foundry facing. Prentiss manufactured about 1.5 million bolts per year.

Perhaps the earliest and most important industrial center in Saratoga County was established in the town of Milton. Hezekiah Middlebrook constructed a dam across the Kayaderosseras Creek at the northern extreme of Ballston Spa in 1830 that created a large pond and valuable power source. A gristmill (called the Blue Mill) was built near the dam that would contain various industries for almost a century. Cotton mills sprang up nearby. A woolen mill was built at the other end of Blue Mill Pond that operated as such for about sixty years before being rebuilt as the county's only chemical pulp mill. An oilcloth factory was established on Bath Street in 1850 that remained in operation for twenty-five years.

Just north of Ballston Spa along the Kayaderosseras, Isaiah Blood established a scythe shop. This small business was an ongoing success that grew in proportions year by year until it occupied a large stretch of the creek and supported several hundred people. Farther up the creek was a hamlet known as Craneville. Lindley Murray Crane relocated there from Massachusetts and rebuilt an old mill to make some of the earliest paper collars and cuffs in the country. His patented process for manufacturing waterproof paper was put to good use until his death in 1874, when the mill became part of George West's paper bag empire. A mile upstream, Milton Center had a strategic location where the Kayaderosseras crossed Middle Line Road, a path so old that the Indians once used it for their annual summer hunting and fishing expeditions. Here was situated a wheel factory, gristmill, extensive tannery, two hotels and several hundred residents. Despite the settlement's prosperity and great promise, it declined rapidly after the tannery burned down and was relocated to Ballston Spa.

Farther upstream was the hamlet of West Milton, a consolidation of Clute's Corners and Speir's Corners. A gristmill was built around 1800 on the northern extreme of the settlement that was later renovated as an important paper mill called the Pioneer. It was one of the county's most successful paper mills in the 1860s, but ownership disputes led to its destruction by arson. It was later rebuilt by George West and was one of the last surviving paper mills in town.

Perhaps the most successful of the industrial centers along the Kayaderosseras was Rock City Falls. Chauncey Kilmer was instrumental in developing the creek in that section by building a paper mill in 1840 at the western side of the hamlet. Later he rebuilt an old gristmill a short distance downriver into a paper mill that was only the second in the country to manufacture paper from straw. This enterprise was operated by him for many years and was known as Kilmer's Stone Mill. Two other paper mills, the Empire and Excelsior, were operated by George West for many years in the production of paper bags.

Other diverse manufactories in Ballston Spa included a soap factory, a paper collar factory, several sash and blind shops, a hoop skirt factory, several large sawmills, a water wheel shop, three carriage shops, two gristmills and, at various times, several mineral water bottling works. In the 1900s, these industries were replaced by several large textile factories, most prominently the Ackshand glove factory and the Ballston-Stillwater complex for the manufacture of knit socks. These gave employment to over one thousand people for decades. The continued prosperity of the tannery and the addition of a chocolate factory at the former Union paper mill allowed the citizens of Ballston Spa to weather the Great Depression much easier than most of the rest of the country. Tufflite Plastics, which manufactured Styrofoam containers for holding flowers and plants, operated for fifty years until 1997 in the old Ballston Refrigerating Storage building. Ballston-Stillwater Knitting also lasted until the 1990s.

The successful and long-lived industry of Edinburgh may be the most underappreciated industrial center in Saratoga County. This is partly because nearly all of the industry was destroyed before 1900 and partly because the physical sites of these enterprises are now mostly under water, so "reminiscing" about them is difficult at best. The first settlers in the town of Edinburgh were farmers who made use of the fertile lands on the Sacandaga River flats. As the population grew, saw- and gristmills were established to serve the several hamlets that were slowly being formed. The abundant supply of trees and suitable sites on the Sacandaga to harness water power prompted several businessmen to take up residence and create a multitude of industrial interests there. Nearly all of the industry exploited the seemingly endless supplies of lumber and took the form of woodenware factories. There were four woodenware factories in Batchellerville: the Batcheller and Jenkins mill, which produced barrel covers, clothespins, bowls, pails, grain measures and small tables, with fifty employees; King, Pearson and Company, which

A load of rakes is being delivered to market from Sumner's Sawmill and Hay Rake Factory. *Courtesy of the Edinburgh Town Historian.*

manufactured butter pails, tubs and a small container called a "kit," with ten employees; Lucien DeGolia's factory, which manufactured hundreds of thousands of his patent washboards, with ten men; and John Millet's woodshop, which contained a planer, lathe and drill press for repairing the machines of the nearby woodenware factories.

Although Edinburgh possessed the greatest number of woodenware factories in the county, it did not have a monopoly on them. One of the largest of these factories was located in Hadley at the northernmost corner of Saratoga County. About one-third of the town's land was suitable for farming, while the rest consisted of heavy forests, making a lumber-based economy only natural. Most industrial activity took place in Conklingville, but the hamlet of Hadley hosted two large paper mills in the late 1800s—the Sacandaga Pulp mills and the Union Bag and Paper Company mills. The latter was established in 1878 by Charles Rockwell, who operated the Rockwell Falls Fiber Company. He later formed a partnership with George West of Ballston Spa to enlarge the mill and equip it to manufacture manila paper. For a few years it was the largest manila paper mill in the country and employed more than seventy-five men.

The hamlet of Conklingville was unique in that it occupied both sides of the Sacandaga River and spilled into two towns: Hadley and Day. The bulk of the manufacturing interests were located on the Hadley side of the border. It was first developed commercially in 1848 by Gurdon Conkling, who built a tannery that eventually became one of the largest in the county. Another large factory was the woodenware factory of the W. and B. Van Vranken Company. This was a significant enterprise, employing one hundred men and consisting of a shop three stories high, two hundred feet long and fifty feet wide.

The largest industry of Clifton Park was derived from beneath the earth's surface—the exportation of molding sand. The iron and steel industries of Waterford and Troy grew in size and importance during the nineteenth century, which required a special type of fine sand to create the molds for such items as stoves, railroad rails and horseshoes. This sand was found in several places around Clifton Park and was easily loaded onto Erie Canal barges for the trip south to Troy and New York City. It is acknowledged that the Albany and Troy foundry business could not have been as successful as it was without the ready supply of Clifton Park's molding sand. The New York State Museum's 1909 Annual Report described the local sand industry:

> *In New York State there are two main areas in which good molding sand beds abound: on the lands bordering the Hudson River from Orange County*

> *northward on both sides of the river to Saratoga County, and in Erie County. The sand is found in shallow beds immediately underlying the sod and often covering many acres. Beds eight inches or more in thickness are worked profitably, a one foot bed over one acre yielding about 1,200 tons. The large increase in business in the iron and steel plants was directly responsible for an increase in the production of molding sand. The production in 1909 amounted to 468,609 short tons valued at $437,402. Of the total production the Hudson River region furnished 450,989 tons or about 96 percent of the total.*

A New York–based company called Whitehead Brothers mined sand from different parts of the state, including Saratoga County. A large boardinghouse was built for its employees, many of whom were Polish immigrants, on Van Vranken Road. A barn on the property held about eighty horses that were used to bring the sand to the canal for loading onto barges. The company had its own dock at the end of Van Vranken Road where 250 tons of sand were loaded onto each boat, which required two horses to pull. The boats had room for two more horses and cramped living quarters for the crew. The Whiteheads built new docks on the Mohawk River when parts of the Erie Canal were closed in 1917 to make way for the new Barge Canal. During the winter months when the canal was drained, sand continued to be mined and delivered near the river in huge piles of varying grades. Later, the sand would be loaded into boxcars and delivered via the railroad. Men worked an average of ten hours per day and were paid fourteen dollars a month plus room and board. After the sand beds in the Albany area were exhausted in the early 1900s, the sand from Clifton Park took on even more importance. A total of nine sand facilities existed at one point in Saratoga County. The industry died out in the middle of the century, just as the Northway was being constructed through town.

South Glens Falls is a relatively "new" village when compared to other villages in the county, with significant settlement only occurring in the mid-1800s. However, the location of South Glens Falls along the Hudson River made it an attractive area for industry once it was established. The Delaware and Hudson Railroad entered the town from Northumberland and headed for the Hudson River on the eastern town border. The successful Glens Falls Paper Company was the largest manufacturer in the town of Moreau. The original paper mill was constructed on the banks of the Hudson River in 1865, just as other paper mills in the county were springing up to take advantage of the high price of cotton after the Civil War. The main building

For a few years, the Hadley Mill was the largest manila paper mill in the country. *Courtesy of the Chris Morley Collection.*

contained almost 25,000 square feet of space and had a smokestack 135 feet high. Another building two stories high and 264 feet long contained the engine room, machine room and vats for water and pulp in different stages of preparation. The Union Bag and Paper Company built a large paper mill in the hamlet of Fenimore, across the river from Hudson Falls, which was later purchased and operated for many years by the Chase Bag Company. The other large manufacturer in the village was the Morgan Lumber Company, which owned four sawmills for different types of wood work. The sawmills had eighteen floodgates on the Hudson River and a sawing capacity of 150,000 standard logs per year (amounting to 80 million feet). This large enterprise at times employed two hundred men, used fifteen canalboats to ship its goods to market and owned forty-five thousand acres of woodland. The Morgan Lime Company owned four limekilns and shipped some eighty thousand barrels of lime in the late 1800s.

The first site of industrial activity in the town of Charlton was on the Alplaus Kill that traverses the southern portion of the town from the Schenectady County line to the Ballston town line. A remarkable number of mills were operated along the Alplaus during the late 1700s and early 1800s. Operating within a few blocks of one another were William Ely's leather factory, Justus Raymond's tannery and bark mill, the Charlton Union Cheese and Butter Factory and Henry Springer's pump factory. A small hamlet called Little Troy showed some promise for growth in the south end of Charlton beginning in 1820. A fulling mill, sawmill, gristmill, carding mill and blacksmith shop were all built in just ten years. The fulling mill supplied clean linen cloth to the collar factories of Troy—hence the name "Little Troy." However, by 1870 Little Troy had become all but abandoned in favor of nearby villages with greater potential.

The "Queen of the Spas" was not particularly known for its industrial output, but the position of Saratoga Springs as the county's largest and wealthiest population center necessarily had a variety of manufacturers throughout its history. These were nearly all considered to be "clean" industries, as large smokestacks, foul odors and loud noises are not conducive to a tourist destination. By far the most famous industrial activity involved the processing, bottling and shipping of mineral water to all parts of the country. Heavy industry was discouraged within the city due to the sensitivity of the tourist trade, but a wide variety of light manufacturing has been carried on over the years. These included the road and vehicle equipment developer Tarrant Manufacturing Company, the medicine firm of G.F. Harvey and Company, the Baker and Shevlin foundry and the Van Raalte textile mill.

Archaeological evidence near Fish Creek suggests that Native Americans lived and hunted in the town of Saratoga more than one thousand years ago. The town also had one of the oldest settlements in the county, with some thirty families making up what was known as Fort Saratoga in the early 1700s. However, a permanent industrial center could not be developed until after the Revolutionary War had ended. Once the war was over and the threat from Indian attacks greatly subsided, the small settlement of Schuylerville expanded into a thriving commercial center. Johannes Schuyler, the village's namesake, developed industry before the Revolutionary War on the lower and upper falls of Fish Creek. His grandson General Philip Schuyler partially developed the falls of Fish Creek to power his grist-, wool, fulling and sawmills. Philip's grandson, also named Philip, built the second cotton mill in the state in 1828, called the Horicon. This mill was purchased by the Victory Manufacturing Company as it expanded its cotton-making capacity. Other mills in the vicinity before 1820 included a fulling mill, distillery, woolen mill and woodenware factory that produced bowls, rakes, axe helves and tool handles. Industry took hold in earnest once the Fitchburg Railroad (later the Boston and Maine) established itself in the village as an efficient distribution system. In 1898, several businessmen established the Liberty

Wall Paper Company a short distance north of Schuylerville that operated for about twenty-five years. On the outskirts of the village near the canal was the Schuylerville Paper Company, established by Daniel Bullard and several partners in 1863 with an investment of $150,000. Within a short time, the mill was turning out four tons of news and book paper per day and employed seventy men and women.

Although Stillwater was located on the Hudson River, the slow-moving waters at that point precluded the development of heavy industry until the mid-1800s. A series of fires further hampered growth, but nevertheless the village has been noted for its various manufacturing plants. The first significant mills, including saw-, grist- and clothing mills, were operated by the prominent Schuyler family. Several of these burned down in 1817 but were rebuilt by Philip Schuyler. A wooden bridge constructed across the Hudson brought business from the east side of the river that had previously gone to Troy and Fort Edward. William Mosher and Elihu Allen constructed a wallpaper mill in 1847 that employed about twelve men. The buildings burned down but were rebuilt by Mosher, Haight and Company. Four large engines drove the wallpaper machine, which turned out one and a half tons of paper per day. After the mill had been in operation for a few years, other capitalists became attracted to the village. Gardner Howland and Sons built

The Saratoga Knitting Mill of Stillwater. *Courtesy of the Stillwater Town Historian.*

Interior view of the Liberty Wall Paper Company of Schuylerville. *Courtesy of the New York State Archives.*

a large paper mill about halfway between Mechanicville and Stillwater and turned out three tons of printing paper per day. Brothers Daniel and William Pemble established a strawboard mill for manufacturing one ton of cardboard per day with a dozen men. The machinery included two large engines, one thirty-six-inch cylinder machine and a Dodge's Patent Dryer measuring eight feet in diameter and eighteen feet long. The Pemble Mills used seven beating machines, two rotary boilers and two wet machines to manufacture three tons of leather board per day in the early 1900s.

The primary industry in the town of Day was that of lumbering. The poor soil conditions found throughout much of the town prevented the development of significant agriculture, but the seemingly limitless supply of timber was enough to attract several hundred hardy residents from an early date. Eliphaz Day, for whom the town was named, was one of the earliest and most successful lumbermen. In 1825, he partnered with Abner Wait and John Johnson to build a dam across the Sacandaga and a larger sawmill. However, since early settlers were occupied with lumbering, no significant hamlets formed for much of the 1800s. A few settlements, such as Day Center, West Day, Croweville and Conklingville, grew to a few dozen people but could not support heavy industry. During the 1860s, the town's population grew as demand from paper mills and tanneries located in nearby villages skyrocketed. A tannery was established by Henry Poor and Son that produced 140 tons of sole leather per year using eight thousand hides and 1,500 cords of wood.

The sparsely populated town of Galway had more than a half dozen churches in the early 1800s but very little industry. A lack of large rivers or a railroad connection stifled its growth, although by 1855 it had six sawmills, two gristmills, two broom handle factories and eight blacksmiths. The largest manufacturer in town was the Mosher plow factory, which featured the patented plows of Israel Mosher. The Seabury family operated carding and sawmills that later became the Seabury Vinegar Works, located near West Corners. The Hudson and French Pottery Works, operated by Nathanial Hodgson (who went by the name "Hudson") and George French, did an extensive pottery business within the village for many years. The 1860 census lists Hodgson's property as thirty thousand drain tiles, four thousand butter pots and jugs, forty tons of clay and twelve tons of stoneware clay. The Stilwell lime and cement factory operated just to the north on Parkis Mills Road. The Parkis mill complex of central Galway was listed in the 1855 census as having "three run of stones, one employee, and works up about six thousand bushels of rye and corn."

In Greenfield, a man named Gershom Morehouse was purportedly the first settler to open a manufactory in what would soon become known as Jamesville in 1786, having established a sawmill there and a gristmill in Milton Center. His sawmill and a second gristmill were the first two such establishments in the town. The settlement remained small and peaceful for several decades until a successful entrepreneur and businessman named John James arrived from New York City in 1836. James was one of the first people to recognize the valuable potential of the twisting Kayaderosseras that flowed through Greenfield and into Milton. He built a paper mill on the creek and installed modern machinery, making it the first paper mill in the county to use machinery rather than hand methods. A large tract of land was purchased on which to build houses, stores, a post office and a bank. A substantial glass factory was built by the Granger family near Lake Desolation to serve Saratoga's spring water industry.

Halfmoon is one of the few towns in the county that did not have a central population center or village once the city of Mechanicville was set off from it in 1910. The presence of a sawmill near Crescent and a gristmill to the north in Ushers suggests the existence of several stable settlements in the mid-1700s. William Peters constructed a combination sawmill, gristmill and plaster mill on present-day Vischer Ferry Road that was known for its attached stone arch bridge over the nearby stream. A Crescent resident named Alfred Noxon did much for the economy of Halfmoon during the middle part of the century by developing a hotel, bank, stores and the Crescent Iron Foundry, which manufactured Franklin Stoves.

To the north of Crescent along present-day Farm-to-Market Road was the famous Anthony Spice Mill, which was successfully operated by Joshua Anthony for over fifty years. He began by establishing a small baking powder business and building it into a sizeable manufactory with a variety of products. In 1882, he received a twenty-five-year contract to supply the Great Atlantic and Pacific Tea Company (better known as A&P) with his baking powder, which was then distributed throughout the country. Anthony added spice grinding and fluid extracts used to flavor cooked food to his product line. The company claimed to produce three thousand pounds of pepper per day. Later he added cornstarch, oatmeal, soda and celery salt. By that time, the baking powder factory was housed in a forty- by sixty-foot, two-story frame building equipped with the latest machinery, while the spice mill occupied a brick addition measuring thirty-six by sixty-six feet. Anthony had a private telegraph line installed at his house near the new Ushers railroad station that connected with his factory and Clifton Park so

This gristmill on Main Street in Middle Grove survived well into the 1900s. *Courtesy of Paul and Marilyn Pastore Collection.*

The Anthony Spice Mill along Farm-to-Market Road in the town of Halfmoon as it appeared in the 1890s. *Courtesy of the Town of Halfmoon Historian's Collection.*

that he could instantly communicate with the cities of New York and other states. He also contributed to the formation of a post office at Ushers. The mill finally closed down in 1920, and Joshua Anthony lived a comfortable retirement until his death twelve years later.

It is believed that the first settlement in the town of Malta took place at Maltaville just above Round Lake, which probably took its name from a malt brewery. Other settlements at East Line and near Saratoga Lake were established before 1800, although the town did not have a larger population center than Round Lake. Malta's only incorporated village never developed any industry, since Round Lake was designed as a religious gathering spot that was not to be disturbed by manufacturing enterprises. Maltaville contained the most industry in town, growing to a peak population of about one hundred people in 1860. The usual saw-, grist- and cider mills served nearby residents. Only the woolen mill did business outside of the hamlet—people from as far away as Greenfield reportedly brought their wool to the mill for processing.

The town of Northumberland was favored with a steam railroad line on its west side and the Champlain Canal and Hudson River on its east side. Early settlers found the area to have abundant farmlands, forests to provide lumber and a variety of streams for power. Several prominent families made the town their home, such as the Gansevoorts and the Billings, who operated numerous businesses. General Peter Gansevoort purchased 1,800 acres of land after the Revolutionary War and built a sawmill, gristmill, woolen mill and flax mill. Jesse Billings Jr., the son of a prominent doctor, made a substantial fortune building canalboats in Bacon Hill, sometimes having four under construction at once. He also operated a general store and had an extensive ice business, in which he sold harvested ice to the Capital Region and New York City. His docks on the Champlain Canal enabled him to ship ice and potatoes to the South and receive supplies for his store and boat building business.

The town of Providence, bounded by Galway to the south and Edinburgh to the north, has historically derived most of its industrial pursuits from its forests. An 1866 atlas of Providence revealed that there were over twenty sawmills scattered around town. Most of the harvested lumber was sold in either the local woodenware factories or in the paper mills of Milton, while hemlock bark was sold to local tanneries. Thomas Shankland built one of the county's first saw- and gristmills at what was known as Whiteside's Corners. Around the turn of the century his land was sold to Jonathan Hagadorn, who came from Galway to establish additional mills and operate a small farm. Because he did so much for its development, the hamlet was renamed Hagadorn Mills in his honor. Between Hagadorn Mills and Fayville were the Trevett family chair factories. The Trevetts moved from Newport, Rhode Island, in 1804 and settled in the eastern part of town near Frenchman Creek. By 1850, Henry Trevett was manufacturing wooden tubs and pails, while his brother Edward had a chair factory that reported making three thousand chairs per year. Ten years later, Henry purchased the chair factory of Fox and Lyon and began manufacturing chairs and wooden rakes. Several more chair shops were purchased in the area, as well as in Broadalbin (Fulton County), by the Trevetts throughout the late 1800s as the business passed from one family member to another. Edward's factory was the largest, utilizing two water wheels to run slitting, boring, mandrel, tapering and circular saws to manufacture more than six thousand chairs per year by 1870. The shop was rather extensive, consisting of a spacious two-story building with separate departments for sawing, assembling, gluing and painting chairs.

As can be seen, ample forests and rivers provided the ingredients needed to establish industrial centers throughout the county. At various times in the late 1800s, the combined output of the giant paper mills in Corinth, Mechanicville, Ballston Spa and Schuylerville exceeded any county in the nation with the exception of one or two in Massachusetts. Related industries such as woodenware factories, tanneries and sawmills added hundreds of more jobs to the several thousand created by the paper mills. In addition, rare industries such as spring water, fire engines, valves, linen thread, mining and ice harvesting gave the county a variety that few others in the country could match.

Nature's Gift

The Mineral Springs

The combination of factors necessary to create accessible mineral springs makes them rare in the world. Only Germany contained springs of greater fame than Saratoga in the 1800s. Many parts of the county were blessed by a wide variety of mineral springs, including Saratoga Springs, Ballston Spa, Quaker Springs and Round Lake. The Saratoga Springs area, in particular, contains dozens of these gifts from nature, many of which were at one time bottled and sold throughout the country.

The most famous spring is generally considered to be the Congress Spring, discovered by a group of distinguished visitors to Saratoga in 1792. John Taylor Gilman had been a member of the Continental Congress in Philadelphia, and since he was the man who first laid eyes on the spring flowing from a crevice in the rocks, someone from his party immediately named it the Congress Spring. In 1802, a young man named Gideon Putnam took a chance and purchased three hundred acres of land surrounding the spring and began building a hotel nearby. After Putnam died, John Clarke purchased the land and began bottling the highly acclaimed mineral water. The spring was encased in a classic Greek Doric structure and surrounded by an elegant park. A bottling plant was built across the street, and in 1823, the first mineral water bottles were shipped to eager buyers across the country. A new and valuable industry was born.

Water taken from the spring was immediately pumped into bottles made by the Granger bottle factory in nearby Greenfield. As soon as a bottle was filled, it would be corked and wired shut to keep in the carbonic gases before being packed into cases for shipment. Since the spring was burdened with

Some of the largest hotels in the world, such as the Grand Union, were located in Saratoga Springs. *Courtesy of the Library of Congress.*

supplying twelve thousand visitors with mineral water during the summer tourist season, bottling operations only took place from November to June of each year. Another anomaly in the spring water business was that all of the commerce associated with it occurred in cities other than Saratoga, since the springs were free and open to the public.

The Empire Spring was likely discovered by Jacob Walton in 1793 and for fifty years afterward was known as the Walton Spring. The Robinson brothers undertook its development in 1846, boring twelve feet down into calciferous rock before finding the source. George Weston purchased the spring two years later, erected a handsome pavilion over it and set up a bottling house a short distance away. In 1865, a wealthy paper manufacturer from Rock City Falls purchased both the Empire and Congress Springs and formed the Congress and Empire Spring Company. Within ten years, it was the largest bottling company in the world, shipping over one million bottles annually. The two springs later became separate entities once again when the corporation was dissolved.

Merchants John and Ziba Taylor decided to expand their operations by building a saw- and gristmill on a new parcel of property about one mile east of the center of Saratoga Village in 1794. Ten years would pass before one of them discovered a mineral spring on the property. Much to their surprise, further exploration uncovered nine more springs. However, tourists were unwilling to make the "long" trip to the Ten Springs area when so many other mineral springs were within easy reach. It wasn't until the Lawrence family purchased the springs in 1859 that they were properly developed and marketed. Caroline Lawrence introduced a new Excelsior Water trademark, pioneered the use of shipping the water in a new type of barrel and developed an improved method of bottling the water using an air pump that she claimed was able to preserve the natural carbonic acid gas. Bottling operations continued into the late 1880s.

There were a large number of other spring companies that operated in Saratoga throughout the century, some of which were long-lived and successful, while others were speculative and only lasted a short while. The Geyser Spring was America's first spouting spring, discovered inside a nut and bolt factory near the Geyser Lake area. By 1875, the guest register had recorded 150,000 visitors to the spring and its bottling house. Various proprietors, including the family of John Clarke, operated the High Rock until 1911, when it was sold to the city to become part of High Rock Park. Hyperion Spring water was cleverly sold as "American Kissingen Water" to foreign-born Europeans who had fond memories of the famous Kissingen Spring of Bavaria. The Seltzer Spring is believed to be one of only two sources for natural seltzer water in the world (the Nassau Spring in Germany is the only other rival). Dr. Josiah Haskins erected a huge bottling plant near the spring in 1868 and even invented a glass tube that allowed visitors to watch the bubbling water as it gushed out of the ground. The Star Spring Company—whose spring was formerly known variously as the Gunpowder, Iodine and President Spring—was one of the first to use block tin–lined oak barrels to transport its water in order to maintain the flavoring and natural gases for long periods of time. The highly successful Saratoga Vichy Spring Company won an award for excellence at the International Exhibition in Chicago and a gold medal at the 1904 Universal Exposition at St. Louis. The French government brought a lawsuit against the company over its use of the word "Vichy," but the Saratoga company prevailed because it had been using the name for a quarter century before the suit was filed.

Saratoga was not the only area to possess a concentration of springs. Ballston Spa, located six miles south of Saratoga, was briefly known as

"America's First Watering Place" after mineral springs were discovered in 1771. New York City entrepreneur Nicholas Low visited Ballston at the turn of the century and decided to establish a tourist center there. In 1804, he built a 150-bed hotel called the Sans Souci and laid out streets in relation to it. Tourists quickly began flocking there in such numbers that several more hotels were built, and the village of Ballston Spa was incorporated. Unfortunately, several of the springs failed just as Saratoga was gaining popularity as a tourist destination, and Ballston Spa became more known for its industrial enterprises than its springs.

A few decades later, when new methods for boring into the earth and properly tubing springs had been introduced, some springs were either redeveloped or newly discovered. The tourists were not likely to return, but the unique properties of several springs prompted businessmen to turn them into successful bottling operations similar to those of Saratoga. One such operation began in 1865 as a local sensation when excavators digging a ditch at the Parent brothers' tannery discovered a substance that proved to be

Indian allies of the British brought an ailing Sir William Johnson in 1767 to the High Rock Spring, one of the county's oldest springs. *Courtesy of the Library of Congress.*

The Hide Franklin Spring was housed in a unique stone building on Malta Avenue, Ballston Spa, measuring eighty feet in diameter. *Courtesy of the Chris Morley Collection.*

a higher grade of oil than what was found in the Pennsylvania oil region. Prospectors purchased the property and began drilling. Although only small quantities of oil were found, the men conducting the boring operation struck a vein of water that shot fifty feet into the air and continued spouting for two hours. The water was tested and proclaimed to be much stronger than any in Saratoga and of "great medicinal value." The Artesian Spring Company was quickly formed, and a handsome brick building was constructed over the spring. When it was later found that the water also contained 7.7 grains of the extremely rare bicarbonate of Lithia, the owners conducted a national marketing campaign that showcased the fact.

Around the same time that the Artesian spring was being developed, a man named Samuel Hides, who was a member of the Spiritualist Society, allegedly contacted the spirit of Benjamin Franklin, who advised him to drill for a new spring on Malta Avenue that would be "for the healing of nations." True to Franklin's word, Hides found a robust spring at 715 feet that spouted 50 feet in the air and had a sustained flow of three hundred gallons per hour. Hides named it the Franklin Spring, and when the water was tested, it proved to have the highest mineral content ever recorded. The Franklin Spring Company began bottling operations in 1868 and continued well into the 1900s under Abijah Comstock.

The Society of Friends, better known as the Quakers, established a community called Quaker Springs a few miles east of Saratoga Springs in 1823. According to tradition, Native Americans used the mineral springs found there for medicinal purposes, and since they were located close to the Hudson River, English and French explorers had known of the springs many years before those of Saratoga and Ballston were discovered. Of the four mineral springs located in the hamlet, at least one, the Paradise Spring of Albert Crook, was used as a commercial bottling operation in the mid-1850s.

Several industries were established in the county to support the mineral spring business. The most famous of these was the bottle-making industry of Mount Pleasant. In addition, a small cooperage industry was established by John McNamara, who moved from Waterford to Saratoga Springs in 1870. A wave of yellow fever was spreading across the southern states at the time, so Saratoga and Ballston spring water was much in demand. McNamara's Waterford cooperage manufactured barrels for the flour industry, but he realized that he could derive more revenue by making barrels for spring water. He therefore purchased a shop from George Brackett and immediately began making barrels for such springs as the High Rock, Star, Geyser, Vichy, Triton and Sans Souci. Each barrel was made by hand of heavy oak imported from Michigan and lined with pure block tin to prevent rusting. Two openings were made in the top: one that contained a pipe that extended to the bottom of the barrel to add and remove water, and another that was secured with a short section of pipe for ventilation. The barrel was filled by connecting the long barrel pipe to the spring outlet tube. As the water flowed, the vent was plugged so that each barrel was shipped under the same pressure as the underground spring itself. The water would be removed by connecting the long pipe to a block tin service line and ice chest cooler. An air pump forced air into the barrel, which in turn forced the water to come out through the pipe. The empty barrels were then returned to the spring to be refilled.

A separate but related industry arose in Saratoga Springs when it was discovered that large quantities of naturally occurring carbonic acid gas could be found in underground caverns. These gases were largely responsible for the continuous flow of water from several springs, such as the Hathorn and the High Rock. Saratoga was the only place in the country in which the carbon dioxide gas was liquefied. Once brought to the surface, these liquefied gases were sold to bottlers to give their beverages a distinctive carbonation. The price of carbonic gas was about twenty cents per pound in 1890, high enough for dozens of borings to be made in an effort to extract as much as possible. Unfortunately, geologic conditions limited the supply, and before long it was apparent that the frenzy of drilling activity was affecting the flow of nearby mineral springs. Finally, in 1909, prominent citizens and spring owners lobbied the state legislature to establish the State Mineral Springs Reservation. State Senator Edgar Brackett introduced legislation that enabled New York State to purchase a large swath of land within village limits that included all of the mineral springs.

The mineral spring water industry took another hit at the beginning of the twentieth century when the federal government enacted "truth in advertising" laws that prevented companies from making unsubstantiated claims as to the medicinal value of the water. For example, Congress Spring water was advertised as having "marked success in affections of the liver and kidneys, and for dyspepsia, gout, constipation, and coetaneous diseases." In 1904, the Congress Spring was sold at public auction for $78,500, and in 1911 it was purchased by the City of Saratoga Springs. Most other spring companies closed as the Pure Food and Drug Act went into effect, although some, such as the Hides-Franklin and Saratoga Vichy Companies, remained successful for many more years.

Local Businessmen Were Inventive

In order to remain competitive, business owners were forced to continually invent new machines, processes and improvements. Fortunately, there were many educated and ambitious businessmen who created a number of inventions that enabled Saratoga County to maintain an industrial economy well into the twentieth century. Several patents exist for nearly every type of industry in Saratoga County, especially papermaking, tanning and toolmaking.

The honor of being the most prolific manufacturer-inventor in the county goes to James Baker of Saratoga Springs. He and his partner, George Shevlin, owned and operated the Baker and Shevlin Company foundry on present-day Route 50. The foundry specialized in fabricating papermaking machines during the heyday of the local paper industry in the late 1800s. Baker submitted some three dozen patents from 1868 to 1922, most of them alone, but many were developed with his son Frederick and George Shevlin. Some of these included a "Centrifugal Pulp Screen," "Wood Pulp Grinder," "Paper Stock Refiner," "Shaft for Winding Paper Rolls" and a "Wood Barking Machine." Several of these machines were reported by the local press to be manufactured and sold to local paper mills as well as many around the northeastern United States.

The most inventive family in the county is arguably the Tarrants, who patented over two dozen inventions over multiple generations. Frederick P. Tarrant started it all with his 1873 patent for an improvement on stove pipe thimbles. At that time, his business dealt mainly with plumbing and steam-gas fittings. His son Frederick Jr. and William Tarrant began patenting a number of road

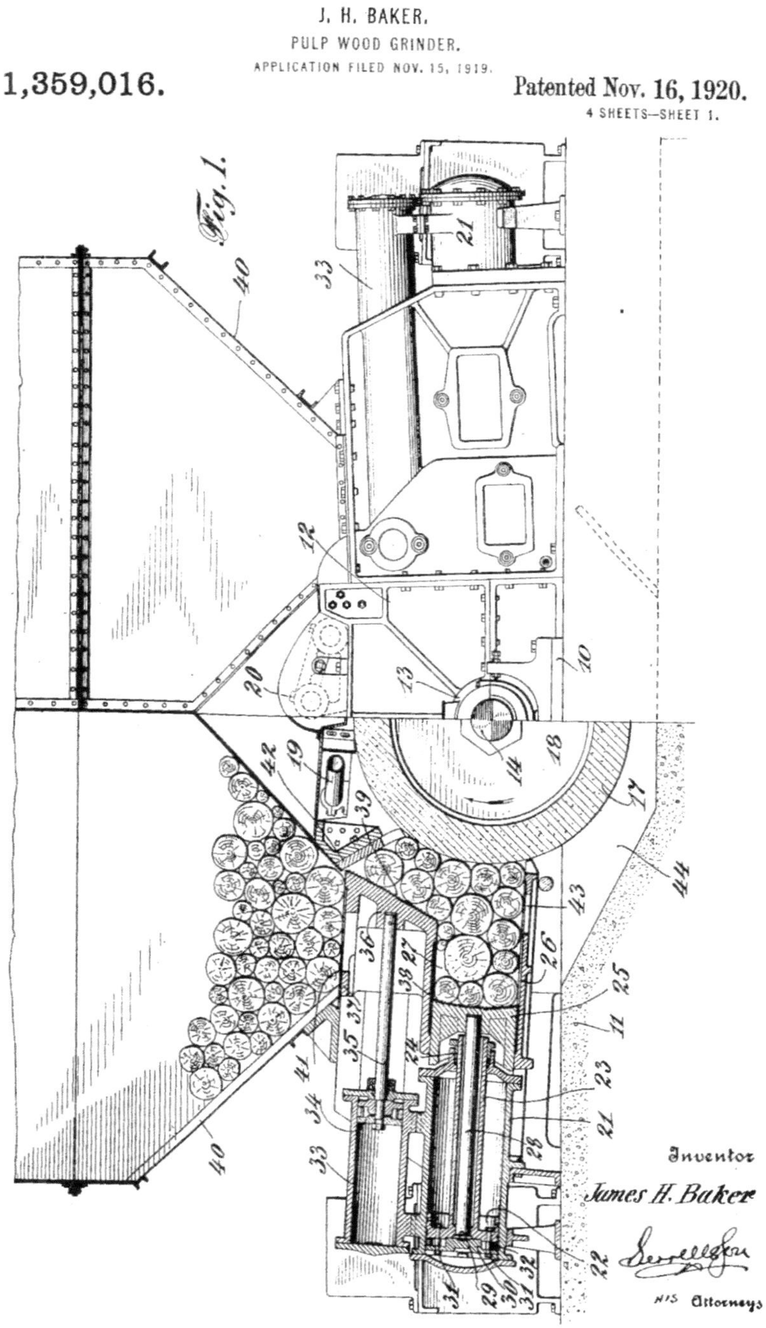

The Baker and Shevlin foundry in Saratoga Springs patented dozens of paper machinery improvements in support of the county's paper industry. *Courtesy of the U.S. Patent Office.*

construction inventions in the early 1900s, and the company slowly expanded to manufacture these new devices. Among the most popular inventions were the asphalt dispensing devices, one of which was the Tarco Scotchman, still manufactured today. Other inventions included street cleaners, a pouring pot, a portable platform and an improved conveyor. As recently as 1992, John Tarrant of the fourth generation patented a "Yard Waste Collection Vehicle."

A variety of other inventions in Saratoga Springs have led directly to the establishment of successful manufactories over the years. Theodore Timby invented and manufactured a globe clock that was sold in such numbers that they can sometimes be found on auction sites today. George Harvey's new machinery and techniques for preparing medicines were instrumental in operating his famous medical supply company. Smith Rawson invented an improved men's suspensory bandage that was sold to the United States military and around the world.

The town of Waterford can lay claim to possessing the most "assigned" patents—those that are specifically signed over to a business—stemming from its long history of successful industry. The Eddy family had a wide range of patents that were put to profitable use in several businesses. The patriarch, Isaac Eddy, invented a process for pulverizing the ingredients in printers' ink that led to a long-lasting business in the form of the Franklin Ink Works. His son Thomas Jefferson Eddy derived a formula for colored ink, while Oliver Eddy invented a new typewriting machine. But George Eddy was by far the most inventive member of his innovative family, with twenty patents and several more unpatented inventions to his name. Most were in support of the Mohawk and Hudson Manufacturing Company foundry that he established in 1847. Such inventions as a chilled railroad car wheel, stove plates and a hollow oven door were all fabricated in his shop. However, the Eddy name did not achieve its greatest fame until he patented and marketed his award-winning Eddy Valve and popular Mohawk Hydrant in the 1870s. His taper-seat, double-gate valve was proven to be one of the most effective water valves on the market and was used by more than half of all the towns in the country. His fire hydrant achieved similar success and is still manufactured today by the Clow Valve Company.

The Button Fire Engine Company of Waterford owed its very existence to the innovations of its namesake, Lysander Button. At a time when hand-operated fire engines were crude and ineffective, Button invented an astonishing number of improvements beginning in the 1830s. Some of his most significant were the "runaround" hose, the first fire engine water pump, the first folding brakes, the first vacuum chamber and the first permanent

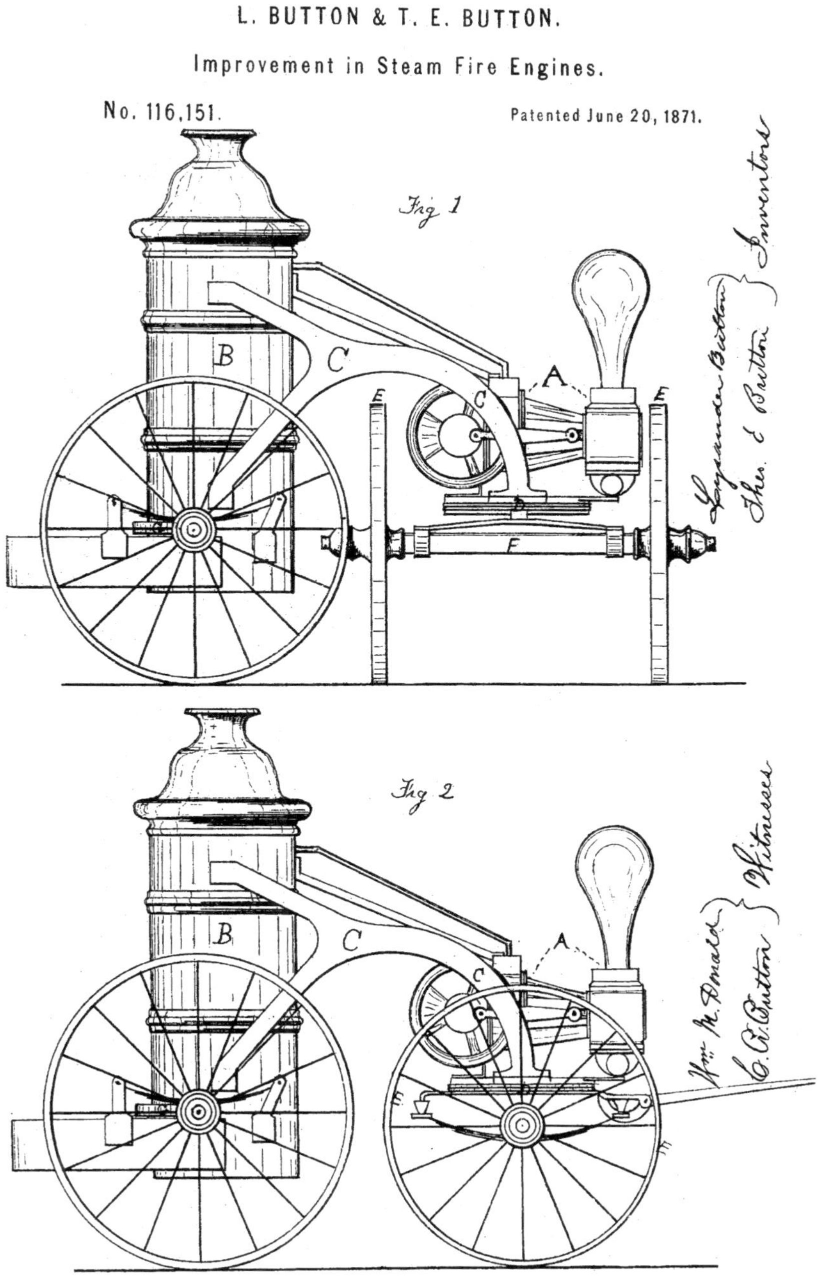

One of Lysander Button's many fire engine patents. This one was developed with his son Theodore. *Courtesy of the U.S. Patent Office.*

suction hose. In 1857, Robert Blake joined the firm, and the two partners began to develop engines that were powered by steam, making improvements to such items as carriage bodies, hose nozzles, hose couplings and pumps.

Dozens of other inventions were developed in Waterford and assigned to local manufacturers. The county's greatest inventor, Henry Miller, filed some one hundred patents during the early 1900s. About one-third related to the textile industry, especially improvements in button hole machines and sewing machines, which were sold to nearby factory owners. Luke Kavanaugh perfected a knitting burr that he used in his knit underwear factory and that also came into wide use around the country. Charles Sulzman, inventor of a strange marine velocipede, also patented a fire hydrant and hydrant valve, both of which were assigned to local businessman John Knickerbacker.

The Hudson River Pulp and Paper Company of Corinth was founded on the strength of the revolutionary Keller-Voelter wood pulp grinder. Although this machine was invented in Germany, the paper mill's superintendent, Warren Curtis, assured the continued success of the mill by developing a variety of paper machines and improved processes. Among these were his "Wood Pulp Machine," "Block Presser for Paper Pulp Mills" and "Apparatus for Separating Fibers and Clay from Liquids." The dominance of the paper mill in the Corinth-Palmer area is illustrated by the fact that over 90 percent of all patents awarded there were paper related. Theodore Elixman invented a lightweight core for winding paper that led to a successful business lasting decades. Richard Bloss patented a machine that removed bark from logs, and Eugene Meurer (a "Subject of the Emperor of Germany") devised two different paper digesters and a pulp screen.

Ballston Spa's robust industry during the 1800s is borne out in the patent record with hundreds of inventions, most in support of manufacturing goods. Benjamin Barber invented a nationally tested water wheel that he constructed and sold around the country. Albert Ackerman invented several lines of gloves that he manufactured at his Ackshand Knitting Company mill for decades. Lindley Crane of the famous Massachusetts Crane family of papermakers developed several types of paper for his mill, including waterproof paper that was among the first in the country to be used in the production of paper collars and cuffs. Both Horace Medbery and Henry Mann, partners in the celebrated Glen Paper Collar Company, patented a multitude of machinery improvements for their factory, including a steam drying wheel, a paper collar cutter and a machine for forming seamless tubes from wood pulp. Ballston Scythe and Axe Works superintendent Ethan Rogers designed and patented new hatchets, axes, tool blanks, a die block and a hydrocarbon burner. The success of Hall, Haight and Company's tannery prompted several employees to patent a variety of leather-making innovations, such as Homer Ely's "Improvement of Preparing Hides and Skins," James Hayward's "Leather Dressing" and Archibald Reid's "Hide Fleshing Machine."

Other towns in the county were well represented by these types of patents. Clifton Park's William Potter derived a formula for creating a roofing composition that gradually evolved into the shingles we use today. During the 1860s and 1870s, Potter manufactured and sold his own shingles before selling the rights to a New Jersey–based company. Lucien Degolia of Batchellerville developed a combination washboard, which he manufactured by the tens of thousands in his washboard factory. John Millet, also of Batchellerville, patented improvements in manufacturing cans and boxes and machinery for producing baskets. Both of these were used in the factory of Benjamin Jenkins. Israel Mosher patented a shovel plow that was manufactured for decades in Mosherville, town of Galway. His former shop survives on Route 29 as a general store.

Despite Mechanicville's large industrial base, only a few related patents were ever filed in that city. Some of the most successful were those of Mayor Homer Eckerson, owner of a bed and mattress factory. A nod to the huge success of the city's brick industry came from George Adams, who patented an improved brick-making machine. Charles and Peter Vredenburgh patented separate brick kilns in 1912 and 1913, respectively. Lewis Noyes supported the paper industry with his machines for coating paper, drying paper and producing lithographic prints.

Schuylerville was another village with a thriving industrial economy that was not well supported by patented inventions. Timothy Cleary, superintendent of manufacturing at the mammoth Liberty Wall Paper Company, patented a machine for producing plush wall coverings, a process for manufacturing wall coverings and a unique design for a wall covering, all assigned to his employer. A scattering of other inventions made an appearance periodically, such as Charles Hermance's "Improved Stone Drilling Machine," John Huested's "Improvement in Stuff Regulators for Paper Machines" and Israel Sonn's "Hosiery and Supporter Thereof."

More than one thousand industry-related patents were filed in the 1800s alone, indicating ongoing efforts to innovate the processes and machinery of the county's factories. This drive died away in the mid-1900s as nearly all industry shut down or moved away. Today, there are still hundreds of patents filed every year in Saratoga County, but the vast majority are assigned to out-of-county companies such as General Electric. These are usually highly technical and developed by teams, with the corporation covering all expenses. The days of the independent inventor-manufacturer appear to be over.

Lumber Drives on the Hudson River

Each year, hundreds of thousands of logs were gathered along the banks of the lakes and tributaries that fed the Hudson River. When spring arrived and the great floods of water from the melted snow packs of the Adirondacks commenced, the piles would be forced into the rapids to form one of the great spectacles of the county, in which floating logs stretched into the distance as far as the eye could see.

The seemingly endless supply of timber along the Hudson River provided an opportunity for lumberjacks to harvest the millions of trees needed to meet the demand for wood in the Capital Region and more distant cities such as New York City and Boston. Finished lumber also supplied the needs of local industries such as paper mills, tanneries, boat builders, coopers, wagon makers and woodenware factories. Even the soap factories in the county relied indirectly on Adirondack lumber, as giant piles of wood ash were collected into "ash sheds" to be subjected to a leaching process, which extracted the lye used to make soap.

The Saratoga County village of South Glens Falls was the main beneficiary of the famous Hudson River lumber drives, as it was located at a point in the river where large sawmills could receive millions of logs per year for processing. The Morgan Lumber Company was the largest of these, producing eighty million feet of finished lumber annually using 250 saws. The company owned forty-five thousand acres of timberland in Saratoga, Fulton and Warren Counties. It also dabbled in the retail trade, operating a shingle mill with the capacity to turn out 250,000 shingles per year and a sizeable box factory that made boxes for several spring water bottling companies in Saratoga.

Although lumbering operations were carried on all year, each stage of the business was performed in a particular season. The year began in late June when lumberjacks entered the forests to cut down hemlock trees for use in local tanneries. Demand for these trees depended upon the market conditions in the leather industry. The men usually built rough-cut log houses near the scene of operations to live in for the season. The hemlock tree harvest continued until July, when the bark would be removed and the trees prepared for sale to sawmills. This "industry within an industry" died away in the early 1900s when most small tanneries shut down and those that remained switched over to using chemicals rather than hemlock bark.

During the fall, other trees would be harvested for processing at the sawmills. The work of hauling trees to the river was done during the winter when it was easiest to drag them through the forest with teams of horses. The logs were collected into piles along the banks that sometimes numbered fifty thousand. It was preferable for the ground to be sloped toward the river so that skidways, or chutes, could be made from the woods to the water's edge. At the bottom of the skidway, the logs were piled up, measured and marked. Once this was done, it remained to wait until the river thawed enough to begin forcing the logs into the water for the trip to South Glens Falls. Because hardwood does not float as easily as softwood, the hardwood trees were generally left standing unless cut for manufacturing establishments such as the woodenware factories of Edinburgh and Hadley. Newspaper reporter John Bulkeley, writing for the *Albany Argus*, described one such Hudson River lumber drive in 1875:

> *The work of floating the logs is attended with no little difficulty and danger. The more rapid the current, the more hazardous it is for the "drivers," viz. the men who follow the logs down the stream. "Jams" are often formed, and the moving mass is frequently stopped by a single log becoming fastened in some hole, or by forming on a rock or bend. Then it is necessary for a man to go upon this floating and treacherous foundation and cut or pry the logs loose. Upon reaching the Hudson River, all logs that have been cut and floated on its tributaries come together, and, forming one grand float, proceed oftentimes with uncontrollable force and velocity to the "Big Boom." Last year, it is estimated that from 1,000,000 to 2,000,000 of logs were in the Hudson, extending from shore to shore, a distance of five miles up the river, and in many places piled upon each other ten or twenty thick, wedged in all conceivable positions, and forming a bridge so firm that teams could cross over with perfect safety. When the "boom" is full, as far as the eye can reach up the river nothing but logs can be seen.*

During a typical log drive on the Hudson River, harvested logs clogged the waterway for miles. *Courtesy of the New York State Archives.*

Upon reaching the sawmills of South Glens Falls and Glens Falls, the river was divided into several channels, each leading to a different mill. Floating platforms allowed workmen to get close enough to sort the logs according to their markings. All logs bearing the markings of a certain sawmill would be nudged into its own channel so that remarkably few of any particular lumberman's trees would be lost in the confusion.

Log driving within a lake was dreaded by the workmen more than in a river. Because lakes have no current, the mass of logs had to be fastened together and driven forward using an anchor, a long cable and an upright windlass. This process, called "kedging," was slow and difficult. It worked best with little wind, so it was generally done at night, although a strong morning wind sometimes drove the boom back to the starting point.

A peculiar custom in this area was to cut logs into lengths of thirteen feet rather than the standard twelve feet used elsewhere. The reason for the

The Hudson River dam log chute in South Glens Falls is clearly visible during low water. *Courtesy of the Library of Congress.*

choice of log length is a mystery, but one theory is that the extra foot was added to the previous standard length of twelve feet to allow for damage on either end of the log during the turbulent journey down the river. Another divergence from custom was to sell logs by count instead of the computed quantity. The standard of count was a log thirteen feet long and nineteen inches in diameter, which was called a "market." Five markets were considered to equal one thousand feet of lumber.

The earliest loggers, observing the immense forests that seemed to stretch into eternity, made little effort to conserve resources or use any type of systematic harvesting program. However, because they were seeking only the largest hemlocks for tanning or pine and spruce trees for lumber, many of the smaller trees and hardwoods were left standing. Some observers in the early 1800s could hardly tell that certain parts of the forest had been harvested during the previous season, as spring foliage camouflaged the winter cut.

The rise of the paper mills in Saratoga County dramatically changed these practices, since wood of all sizes and types could be used in the chemical pulp wood process. As one historian noted, "With the sawmill in view, only the full-grown trees were cut, but with the pulp mill in view, large and small, young and old went down before the axe." Whole swathes of forest were clear cut and hauled to the paper mills using logging railroads or tractors. By the end of the century, there were over two dozen large paper mills in operation within the county, each requiring several cords of wood daily. Only after the paper mills began to close down were the forests allowed to recover. The last log drive of full-length logs occurred in 1924, when logs began to be cut into standard four-foot pulp sticks used by the paper mills. Around the mid-century mark, the advent of improved roads and trucking spelled the end of the Hudson River log drives.

The Cotton Mills

The village of Victory is unique in that it was both created by and named after a manufacturing company. As the Industrial Revolution spread across the country in the early 1800s, businessmen sought out land located near water to power the new machinery. Three wealthy men from Boston named Jared Coffin, Enoch Mudge and David Nevins visited the Fish Creek area where General Philip Schuyler had built flour, linen, fulling and sawmills soon after the Revolutionary War ended. This abundant water power, coupled with the construction of the Champlain Canal and a steam railroad nearby, prompted the men to send Benjamin Losee and Pickham Green to establish a cotton mill just outside of the growing community of Schuylerville.

The Saratoga Victory Manufacturing Company was incorporated in 1846. Its investors spent an amazing $425,000, equivalent to almost $9.5 million today. The large outlay of funds made it the most valuable manufacturing plant in the county for many years. It was a success from the start, and a settlement of hundreds of people rapidly grew up around it. The village of Victory was incorporated under New York State general laws three years later. The name "Victory" was used to recognize the American victory over the British in the famous Revolutionary War battle that was fought nearby. Gates Avenue was established as the main thoroughfare. The expansion of the Victory Mill coincided with the Irish potato famine in the pre–Civil War period, leading to large numbers of Irish immigrants coming to America looking for work. By 1860, about one-third of the total population of the village was Irish born.

Within a year of opening, the mill employed over 350 men and women who operated 12,500 spindles and 309 looms. The principal output was cotton cloth, of which 1.8 million yards were produced a year. The Schuyler family's Horicon Mill was purchased in 1857, and the main factory was greatly enlarged until it measured 274,000 square feet in size. Sawmills and gristmills were acquired in the Grangerville area. The largest sawmill employed 8 men and produced 225,000 finished boards annually.

Despite the respectable number of residents in the village, a full complement of community institutions was never established due to its proximity to Schuylerville. Besides the churches and school, there was a fire company and a few retail businesses, as well as the Victory House and Sheehan's Hotel to accommodate visitors. The post office was opened six years after the mill, with Roland Coffin as its first postmaster. Several local churches were built on lands donated by the company, such as St. Steven's Episcopal Church, the Notre Dame de Lourdes Church and the Church of the Visitation. The four-room schoolhouse was designated as school district #4.

Unlike many other mills, the Victory Mill was built from the start to produce goods for the retail market, and thus, all of its production was exported down the Champlain Canal to ports in Albany. Transporting finished goods to the outside world became much easier with the arrival of the Boston, Hoosac Tunnel and Western Railroad in 1882. A spur was constructed from the Schuylerville depot to the mill itself.

For several decades, the number of employees remained steady at about 250, while the mill had increased its capacity to forty-seven thousand spindles. Cambrics and silesias were the two primary types of cotton products manufactured at the mill. Cambric is defined as a "fine, firm, plain weave balanced fabric with starch, and has a slight luster on one side." Today, it is mainly used as a coating for professional playing cards to make them easier to handle and last longer. Silesia is a "lightweight cotton twill lining with a calendered glaze" used for dress linings and pockets.

A lawsuit brought against the mill by owners of the nearby Bullard paper mill gives some insight as to how the waters of Fish Creek were utilized for power. The Victory Mill dam created a pond of about one acre in size. Just above the dam was a natural ledge of rocks known as "Marshall's Reefs." A bit farther upstream was a similar ledge known as "Winne's Reefs." A series of canals was constructed around the two reefs and regulated the flow of water through them with the use of gates. In times of drought, the gates would be closed each night so that the natural ponds behind the reefs could fill with water. In the morning, the gates would be slowly opened to ensure a

The Horicon Mill was built by the Schuyler family and acquired by the Victory Manufacturing Company in 1857. *Courtesy of the author.*

The Victory Manufacturing Company cotton mill in the village of Victory. *Courtesy of the New York State Archives.*

continuous flow of water during the day. The lawsuit alleged that the Bullard Mill—which operated twenty-four hours a day, like most other paper mills—was sometimes forced to stop its machines in the evening when the Victory Mill closed its gates. The courts decided that the Victory Mill was using its share of the water legally and dismissed the case.

The size of the Victory Manufacturing Company allowed it to remain competitive long after all other cotton mills in the county had shut down, but by the late 1800s, it, too, was struggling. Consolidation within many industries caused the mill to be purchased by the American Manufacturing Company in 1910. James Blacklaw was named resident manager. Prospects for the future seemed bright as the company expanded the plant with a new concrete building. A second, mixed-use brick building was even constructed for the convenience of employees and their families, consisting of a barbershop, bowling alley, restaurant, baths and a community room. By 1925, the village had reached its peak population with over one thousand residents. However, following the trend of most other textile manufacturers, company management decided to move the Victory operation to Alabama to take advantage of less expensive labor costs and access to raw materials, putting two hundred people out of work. The mill's manager summed up the situation well when he stated, "On this class of goods there are very important economies in the south so that the northern mill is too severely handicapped. Because of lower living costs and a milder climate, wages are considerably less. The south is less restricted with regard to hours of labor. Proximity to cotton in the raw state eliminates largely the high rate of cotton now shipped to Victory Mills."

The plant sat idle for nine years until it was purchased by the United Board and Carton Corporation for the manufacture of folding cartons. After several more changes of ownership, including a promising run by the Clevepak Specialty Packaging Company, which employed 120 people, operations ceased in the year 2000. The giant factory buildings are left behind as a reminder of the mill's success in years past.

The Victory Mill was by far the largest cotton mill in the county, but it was not the only one, nor was it the first. Cotton factories are mentioned in the early records of Schuylerville and Stillwater, although these were small and may have only served the needs of the immediate vicinity. A cotton mill with two thousand spindles was built soon after the King's Power Canal was completed in Waterford. However, the enterprise did not last long, and the building was purchased by the King Dye Works.

The cotton factory erected by Nicholas Low and several associates in 1811 is both one of the earliest industrial endeavors in Ballston Spa's history and the most spectacular financial failure. After Low built the Sans Souci Hotel and supervised the plotting of streets and neighborhood blocks for the young settlement, he decided to utilize the Gordon Creek to build what would have been one of the most modern cotton mills in the world. A building 160 feet long and five stories high was constructed of brick and stone on the banks of the newly excavated river, along with a store, office, outbuildings and residences for workers.

It was calculated that the complex had the capacity for eleven thousand spindles and cost $80,000 to build and equip, the equivalent of almost $1 million today. The engine department contained a massive, imported iron beam measuring thirty feet long that was suspended on two iron pillars twenty-five feet in the air. Iron wheels weighing several tons were turned by the shaft, which itself was turned by the latest steam engines, a rarity at that time, since most mills were powered exclusively by water wheels. As the *Saratoga Journal* described it, "You hear the harsh rumble of wheels, iron against iron, and the roar of rollers and spindles, not unlike that of the fall of immense water, yet see nothing like the moving power to all this—no fall or force of water, nothing that the eye discovers gives motion to the unwieldy machinery, but all being moved by a hand unseen, and out of the researches of the eye, very greatly enhances the singularity of the scene."

The massive undertaking was, unfortunately, doomed to failure. Conflict with Great Britain prompted the United States government to enact the Embargo and Non-Importation Act and the Non-Intercourse Act, which effectively blocked the shipment of cotton goods. The War of 1812 prompted the passage of another Non-Importation Act. Nicholas Low was hoping to take advantage of the increase in prices that resulted from the loss of cheap British cotton. However, the war quickly ended, and the embargoes were lifted even as the Ballston Spa Factory was starting up in 1815. To make matters worse, after only a short time in operation, the great iron beam that turned all of the machinery broke and caused extensive damage. Faced with the prospect of rebuilding the plant only to reap much lower profits than anticipated, the difficult decision was made to abandon the enterprise.

Despite the discouraging and expensive setback, Ballston Spa was home to the most extensive collection of cotton mills in Saratoga County. Jonathan Beach and Harvey Chapman first developed the northern part of the village around Blue Mill Pond by erecting a woolen mill near the pond and two cotton mills on the other side of Milton Avenue. Samuel and James Cook purchased and operated the three mills to manufacture cotton cloth and seamless cotton bags. Ziba Cook managed another cotton mill nearby that later became a paper mill, and Hiro Jones used the future paper bag factory as a knitting mill.

Ballston Spa native James Cook was a prominent lawyer and one of the organizers and early presidents of the Ballston Spa Bank. He later attained positions as a state senator, the New York State comptroller and superintendent of the State Banking Department. Cook and his brother Samuel purchased the three cotton mills on the "island" (near Mechanic Street) and ran an extensive business manufacturing "Cook and Company's Seamless Cotton Sacks." By 1850, the company was operating over six thousand spindles and producing 1.9 million yards of cotton cloth. His cotton sacks were "fully warranted not to separate where interwoven" and offered in lots of three hundred or five hundred. The products included ninety-eight-pound and forty-nine-pound bleached or unbleached flour sacks and one-hundred-pound, fifty-pound and twenty-five-pound bleached buckwheat and meal sacks. Special arrangements were made with the Saratoga and Schenectady Railroad to deliver the sacks at reduced prices around the country. The flour sack business was carried on successfully for a number of years until the Civil War disrupted the production of cotton. Cook was finally compelled to close his factories in 1862.

Despite the cotton embargoes and shortages, a Massachusetts businessman named Jonas Hovey purchased nearly all of the mills in Ballston Spa and restarted the idle plants. Although there were some periods of inactivity, he successfully operated his mill complex from 1862 until poor health forced him to retire ten years later. He regularly employed about two hundred people, although this varied as cotton supplies were interrupted. The two main cotton factories were running 9,696 mule spindles, 2,224 frame spindles and 236 looms, which produced about 400,000 pounds of cotton and 3.8 million pounds of cloth. The manufactured cloth was sent to other cities to be bleached and printed. At Hovey's death, the mills lay idle for several years before being purchased by George West, who converted them all to paper and paper bag factories. That left only the Victory Mill as the last surviving cotton mill in the county.

Tanning and Leather-Making

The presence of large amounts of hemlock trees led to the development of the first tanneries, since hemlock bark naturally reacted with the oils of cowhides to accomplish the tanning process. Because it was so much more expensive to transport hemlock bark than it was to transport animal hides, Saratoga County was home to a larger number of tanneries than the other counties in the Capital District combined.

Tanning leather at the first mills was primitive by today's standards. Individuals would drop off cowhides at the factory to be split and initialed with their name for later identification. The hair would be scraped off the hides and put into vats six feet square and four feet deep. Often these would be sunk into the ground for easier access. Alternate layers of hemlock bark and hides were loaded into the vats and covered with water. After soaking for six months, the hides were removed to be washed and dried on racks. Once dry, they would be treated with chemicals and polished to a finished leather. The length of time it took to tan leather obviously limited the amount that could be produced each year, so that tanneries were only capable of supplying the needs of local residents.

Early records mention the presence of many tanneries throughout the county, but the ownership, output and other details of these are scarce. Over the years, dozens of tanneries were established deep in the woods and operated until the local hemlock bark supply was exhausted. The wooden structures have all rotted away long ago, leaving no trace, but a few of the most successful tanneries have at least been mentioned in the early histories. A few of these include the Lyons and Prindle tannery of

Beecher's Hollow, Seth Rugg's tannery in Milton Center, the tannery of Powell and Company in South Corinth and the William Fowler tannery in the Day hamlet of Croweville.

The first large tannery in the county was the Lynwood Tannery, established by Gurdon Conkling in 1848. Conklingville was by far the largest industrial center in the town of Hadley, where a dam was built on the Sacandaga River by Isaac Barber in 1831. This dam was carried away by high water and replaced by Conkling, whose efforts led to the growth of the hamlet that eventually consisted of fifty dwellings and a variety of businesses and manufactories.

The tannery passed through several hands until it was purchased by H. Poor and Sons, a firm that consisted of Henry Poor and his sons Eben, John and Charles of Boston, in 1863, at which time it achieved its greatest production. The main building measured seven hundred by forty feet and stood two stories high. Other buildings consisted of a rolling room, oil house, engine house and storehouse. Five water wheels provided power, supplemented by a thirty-horsepower steam engine. Each year, the tannery used 4,500 cords of hemlock bark harvested from the nearby town of Day, brought down the river by company-owned scows and a small steam tugboat. The hides were imported from Texas, Mexico and South America. Approximately twenty thousand hides were processed each year to produce 350 tons of sole leather with a value of $150,000. Up to forty men were employed year-round.

Compared to the much larger and long-lived tanneries of Ballston Spa and Conklingville, the Morey brothers' tannery of Saratoga Springs received little attention. Unknown to many residents was the fact that the Moreys came from a long line of tanners that stretched back before the Revolutionary War. Nathan Morey, the patriarch of the family, was born in Massachusetts and moved to Troy to establish a tannery. His youngest son, John, was born there and learned the leather trade. He took over the business when the elder Morey died in 1837. He then relocated to Saratoga County and carried on the trade in both Saratoga Springs and Milton. He operated a tannery in Milton Center for several years that would later become the Haight and Company tannery.

John Morey and his wife, Delinda, had two sons who followed in their father's and grandfather's footsteps. Unlike the Bull's Head tannery run by Haight and Company in Ballston Spa, the Morey tannery was not in operation on a consistent basis. When an arsonist set fire to the building in 1867, the tannery had been closed down for the better part of the summer. "They hope to be able to rebuild and go to work soon," read the report, "although the season is so late that they may not be able to start before spring." Fortunately, insurance covered $11,000 of the $17,000 in losses, including $7,000 for building damages, $500 for tools, $150 worth of bark used in the tanning process and $1,500 in leather stock. For some reason, Nathan left the tanning business just a few years after patenting a leather-pressing machine to open a boardinghouse on Franklin Street. His brother Robert carried on the leather business until 1892, when he moved to Saratoga Lake and built the Piscatory Hotel on the western shore.

One of the most sizeable tanneries in the county is almost entirely forgotten and rarely mentioned even in history books that were written when it still existed. The partnership of Dudley and Voorhees first established a tannery in the town of Providence around 1850. Ten years later, it had grown to employ twenty-five men and produce 400,000 sides of shoe sole leather with a value of $75,000 (over half a million dollars today). Henry Sanford assumed ownership and expanded the tannery to its greatest extent. The T-shaped structure measured 280 feet by 66 feet on one wing and 90 feet by 24 feet on the other. A forty-horsepower water wheel powered the bark mill, two rolling mills, two hide mills and four pumps. The community of Glenwild grew up around the tannery and became so large that it soon required a schoolhouse.

Ballston Spa was home to several tanneries from an early date. One that operated for decades under various owners was originally established by Chauncey and Arthur Parent and known as the Parent Brothers Tannery. Some excitement occurred in 1865 when workmen digging a ditch in the rear of the property discovered an oily substance seeping up from the ground. Tests revealed that it was crude oil equal or superior to any other oil well in the country. Excitement spread through the village as oil prospectors erected derricks and bored numerous holes in an effort to find a vein. The Parent brothers sold their land for the tidy sum of $60,000, making them instantly quite wealthy. The new owners did not get their money's worth, however, as no other significant quantities of oil were found.

Once the prospectors realized that there was little oil in the region, the tannery was sold to local businessman Magnus Schwarz. Schwarz began his career in a partnership with Louis Muhlfelder under the name Schwarz and Company. The two opened a general store on the corner of Milton Avenue and Ford Street and operated it successfully for about twenty years. During Schwarz's management of the tannery in the 1880s, it utilized forty-five vats for soaking hides and consumed four hundred cords of hemlock bark per year. Approximately five thousand sides of upper leather were tanned and shipped annually by its five employees. Unlike larger tanneries, however, Schwarz's plant did not have the capability to finish the leather, so it was sold as it dried to be curried elsewhere. The tannery was destroyed by fire in 1886 and shut down.

Although the Lynwood tannery of Conklingville was the largest of its kind in the county for several years, it was supplanted by the tannery of Haight and Company. Haight's original tannery was in Milton Center along the banks of the Kayaderosseras Creek. It was a large business for its time, and a thriving hamlet of several hundred inhabitants was supported by it. The production building was five hundred feet long and from thirty to one hundred feet wide, consisting of a multitude of subsections and additions. The bark mill, boilers and engine were in the center, the tannery vats were to the west, the currying shops were in the rear and the finishing department was in the east end and on the second floor. Haight used nothing but hemlock bark, forgoing the chemicals that other tanneries used to speed up the process. He also didn't lay down the hides flat in the vats but, rather, suspended them so the liquor could penetrate both sides equally. He used about two thousand cords of hemlock per year and finished eight hundred sides per week. The factory employed 65 men in 1875; five years later, it had grown to employ 150 men. The well-kept property was described by the *Ballston Journal* as containing large buildings, "being two stories in height with basement, and are in good repair throughout, their exterior neatly painted, thereby giving the appearance of a first-class situation, as indeed it is. It is the largest manufactory of the kind in the country."

In the winter of 1881, a large fire destroyed most of the tannery buildings. It was discovered by night watchman Charles McCarty on the second story near the elevator. The only other person in the building was engineer James Fitzpatrick, who was just getting up steam for Monday's shift. Within a few minutes, the entire main building was engulfed in flames that could be seen in Ballston Spa five miles away. The stables and office were also burned, but the nearby residences, store, dry houses and Literary Society's hall were all saved. Total monetary losses amounted to a whopping $125,000, but Milton Center would soon lose much more. Rather than rebuild the tannery as it was, Haight decided to move the plant to Ballston Spa. Within just a few years of the tannery's removal, Milton Center had almost become a ghost town, as most of the workers moved closer to Ballston to continue their employment.

The site of the new tannery in Ballston Spa had a long history that stretched back to the era of Nicholas Low, who is credited with establishing Ballston

Spa as a village. He attempted to expand into manufacturing with an ill-fated foray into cotton production, but the enterprise had to be abandoned. For years, the site was vacant until being purchased by John Wait, who used it to produce oilcloth for over two decades. After Wait's death, Blittersdorff and Company purchased the site for making paper boxes. Allen and Heaton, manufacturers of the black hard-stone called emery, then commenced a promising business producing emery wheels, artificial grindstones and free-grit whetstones, but they only operated for a few years.

As each new business was established, the buildings were refurbished or expanded to suit the needs of the new owner. Therefore, when Samuel Haight relocated his tannery, the buildings were already quite large. It was well situated on Bath Street with the business district just a few blocks away and the Delaware and Hudson mainline running along the rear of the property. There was also ample property available for expansion that would soon be put to good use.

The Bull's Head Tannery of Samuel Haight in Ballston Spa (top) was considered to be one of the country's largest during the late 1800s. *Courtesy of the Paul and Marilyn Pastore Collection.*

A view of the "Beam Room," where hides were washed, stripped of hair and soaked in chemicals to prepare for the tanning process. *Courtesy of the Pat Streifert Collection.*

Haight lost no time in rebuilding his enterprise, since his products were in high demand and insurance covered 90 percent of his losses. The plant in Ballston Spa was largely completed within a month of the fire in Milton. The tanning room measured 480 feet long by 45 feet wide. The dressing and finishing department occupied a building four stories high, measuring 200 feet by 50 feet. The bark mill was level with the elevated Delaware and Hudson mainline, and a railroad switch delivered bark directly to it. The bark was then run down a long chute to the boiler house.

The tannery went through several ownership changes during its long life. It was operated by the firm Haight and Company until Samuel Haight's death in 1891. Matthew Vasser sold his interest to Thomas Hall, and thereafter it was known as Hall, Haight and Company. In 1899, the American Hide and Leather Company was organized to consolidate several large tanneries, referred to as the "Upper Leather Trust." Hall became the first president of the trust, succeeded by Theodore Haight and Carl Danner.

Within fifteen years of the move to Ballston Spa, the tannery was one of the largest in the nation, thanks to the efforts of Haight's two sons, H. Vasser and Theodore. The firm was reorganized when it was purchased by

American Hide and Leather. In that year, it employed 350 people and had sales of about $750,000. Over 900 hides were processed each day, creating an output of 1,800 sides of leather. It was an immense complex, comprising over twenty-five buildings and covering several acres of land on both sides of Gordon Creek. It was also highly profitable. The corporate entity reported net earnings of $867,000 in the last quarter of 1917, compared with $1.2 million in the same quarter of 1916. As these figures include all the divisions of American Hide and Leather, it is not known how much profit came from the Ballston division, but the fact that the plant remained in business for so long is evidence that it was one of the corporation's more profitable subsidiaries.

Hides arrived from large meatpacking plants either salted down in bales or pickled in a brine solution to keep them from deteriorating. After they were moved into the factory by a conveyer belt, the hides were put into a giant washing machine capable of holding hundreds of sides at a time to remove the salt or brine. A lime bath would then remove any stray flesh, and another machine would remove the hair. Afterward, the hides would get an acid chemical bath and were then ready for tanning. A mineral solution was used to tan "upper" and pocketbook leather, while a vegetable solution made from sumac was used to preserve sole leather. When the hide was removed from the mineral tanning process, it was officially called leather. After the tanning process, the leather was wrung dry and sorted according to weight, grade and size before being run through a machine that split the leather into a uniform thickness. The main piece, called the "grain," would be dyed to whatever color was ordered. The leftover pieces would be used to line the insides of shoes. It only remained for the leather to be dried, buffed, softened and "plated" (when the pigment was ironed into the leather). There was very little waste in the plant, as the hair removed from the sides was cleaned by a machine and used for upholstery, and the shavings from the edges of the hides were collected to be used as fertilizer.

The tanning industry experienced ups and downs in the twentieth century. For example, changes in the styles of women's shoes coupled with automobile manufacturers using painted cloth and leather substitutes in the 1920s led to a decline in demand for leather products. The company also had to deal with a lack of control over the supply of raw materials, since the rawhides used were a byproduct of the meat industry.

Over the years, the tanning process became more and more mechanized. By the mid-1900s the plant was producing twenty-two thousand sides of leather each week using the "chrome" method, or acid tanning, and employed 325 people. It was said that operations had been so mechanized that the output of 1950 would have taken thousands of men to accomplish in the 1800s.

The tannery was one of the longest-lived in the area, especially when counting its existence for years in Milton Center, and was partly responsible for Ballston Spa's ability to report 100 percent employment several times during the 1900s. However, that dependability came to an end when American Hide and Leather went bankrupt in 1956. Howes Leather Company purchased the plant and renamed it the Ballston Tanning Corporation. The company attempted to maintain operations as before, but changes in the market made it much more difficult to generate profits. The parent corporation also operated a tanning plant in Cincinnati that produced sole leather. The Ballston factory continued to be used for upper leathers on shoes. Prospects were bright for the two plants to merge, along with another in Pennsylvania, and become more competitive with other leather companies. However, Howes Leather was unable to exploit the possibilities of a merger or refine the process for producing upper shoe leather and experienced continued losses. The plant finally closed its doors in 1960, putting one hundred men out of work. The cessation of operations meant an end to over a century and a half of tanning in Saratoga County.

Keeping It Cool

The Ice Business

Until there is a power outage that lasts for more than a few hours, we in the twenty-first century tend to take refrigeration for granted. Until the 1940s, however, people were forced to figure out other ways to keep their food from spoiling during the spring and summer months. That's where the ice harvesters came in.

Harvesting ice was hard work involving long hours toiling under harsh winter conditions. The men were well paid and highly respected for their labors. It took considerable strength and stamina to work the ice saws and pikes (long wooden poles with a steel point and hook on the end) for hours at a time in subzero temperatures. There was also a great deal of risk in the ice industry, as it was largely dependent upon the weather. A warm winter would wreak havoc on the ice business, resulting in a harvest of fewer than 100,000 tons. In other years, the harvest could exceed 4 million tons.

It is believed that the first ice harvesting operations in this area originated in 1825 on the Hudson River near Schuylerville. The river proved to be the most convenient location for a commercial operation because large blocks of ice could be cut and stored in sheds that were constructed along the banks. When the river thawed in the spring, the ice was loaded onto canalboats and hauled by mules down the Champlain Canal to Albany and New York City, where "northern ice" was in high demand.

Boat builder Jesse Billings of Bacon Hill made full use of the Hudson River during the winter months. He had two docks on the river that were used to receive fuel, supplies and lumber and to ship out crops and ice. During a downstate ice famine in 1881, companies from New York City

Ice cutters on the Mohawk River in 1900. *Courtesy of the Town of Clifton Park History Collection, Clifton Park–Halfmoon Public Library.*

staked out ice claims on the river all the way up to the Adirondacks. Billings had an icehouse constructed capable of holding 100,000 pounds of ice. His friends wondered how he could harvest so much ice, since much of the river as far north as Gansevoort was reserved. When the river froze over, Billings had his men cut a wide channel down the center of the river so that he could float large cakes of ice downstream. The companies that had staked out the river filed injunctions against him, but Billings claimed that they could not legally obstruct navigation. He won the battle and was able to reap large profits from the endeavor.

The county's largest commercial ice operation was conducted by the Saratoga Ice Company at Loughberry Lake, located outside of Saratoga Springs. Under the ownership of Leo Grippen in the early 1900s, this company became one of the largest ice harvesting operations in the Northeast. Every winter when the ice attained a thickness of one to two feet, teams of men and horses cleared the lake of snow and began cutting. Tractors were used to drag floats of ice across the lake using a chained conveyor mechanism. The ice was then stored in an old boathouse and in ice sheds that were constructed along the southwest shore. When the sheds were filled to capacity, huge ice cakes were stored outside covered with thick layers of sawdust. As many as six thousand cakes of ice, each weighing three hundred pounds, were stored each year. In the summer, ice would be delivered to local residents or sold over the counter at the company store on Maple Avenue. Prices at that time were about ten cents for twenty-five pounds and a quarter for fifty pounds.

George Fedush purchased the Saratoga Ice Company during the Great Depression and changed its name to Saratoga Ice Distributors. Fedush made use of modern trucking to deliver the ice and some of the first refrigeration units to keep it frozen during the summer. Soon he installed refrigeration machinery at the old waterworks building on Excelsior Avenue and began producing ice year-round. A few years later, the company was purchased by Percy and Charles Dake of Stewarts Shops.

Substantial ice harvesting operations also took place on the placid Mohawk River. The New York Central employed about thirty men during the winter to cut ice into small blocks and load it onto nearby freight cars for delivery to its icehouses. There are several records of lawsuits among the ice harvesting companies over who had the right to certain parts of the Mohawk River and the Mohawk Basin. At the turn of the century, the ice industry on the Mohawk suffered when the City of Schenectady discharged sewage into the river while a new water treatment plant was being constructed, shutting down harvesting operations.

Daniel Foote established a thriving ice harvesting business in 1893. The Sylvan Lake Ice Company was based in a pond on top of a hill overlooking the present-day Brookside Museum in Ballston Spa. The twelve-acre pond is fed by the Gordon Creek and has been a site of industrial activity since 1800. It was first called the Red Mill Pond and then became Westcot's Pond, Willow Lake, Sylvan Lake and finally Foote's Pond. A visitor named James Fennimore Cooper used Mohican Hill, a steep incline on the east side of the pond, as one of the locations in his book *Last of the Mohicans*.

After purchasing the pond and surrounding property, Foote constructed an icehouse and prepared to open his ice harvesting business. At first it was named the Willow Lake Ice Company, but he soon changed it to the Sylvan Lake Ice Company. Horse-drawn wagons ascended the path each winter to the pond, where men would cut out ice cakes for storage in the icehouse and several storage sheds. Sylvan Ice served the business district and residents in the southern end of Ballston Spa, while Elmer Cook, operating out of a pond near Rowland Street and Route 50, served the northern end of town. Although Foote died in 1927, his wife and son carried on the business until 1947, when the property was sold to the American Hide and Leather tannery. By that time, many residents had their own refrigerators and no longer needed ice deliveries.

Ballston Spa was also home to one of the first and largest refrigeration companies in the county. The Ballston Refrigerating Company was established in 1894 in a large brick building straddling the Bath and Low Street block, convenient to the Delaware and Hudson Railroad mainline. The company used high-quality timber, thick insulation and air chambers to keep the cool air inside, while the doors and windows were made of several panes of glass that swung on ball-bearing hinges. The rooms were cooled by fluids passing through a system of pipes on either side, with board curtains placed to maintain equal circulation. Some rooms, such as those used to store fruit, were kept at thirty-two degrees, while others used to store meat were kept at between ten and thirty degrees below zero. Customers included merchants, market-men, fruit producers, agriculturalists and farmers. Even a few wealthy individuals paid the company to store their expensive furs during the summer to prevent insect damage. The company was highly successful for several decades and continued to expand. In 1911, an addition was added to the original building with an entrance on Low Street. However, as refrigeration technology enabled more and more companies and individuals to obtain it themselves, the need for a dedicated business decreased accordingly. Tufflite Plastics purchased part of the building in 1950 and later occupied the rest of it as Ballston Refrigeration reduced operations and finally shut down.

The Mount Pleasant Glass Factory

A small clearing north of Lake Desolation is all that remains of a unique glass factory that operated for about twenty years in the wilderness of Greenfield. Even more unique was the village that was built for the sole purpose of operating this factory and how its residents were later forced to move to Saratoga Springs at the behest of the factory's biggest customer.

The Mount Vernon Glassworks of Syracuse was owned by brothers Oscar and Charles Granger. In 1840, they obtained a twenty-five-year contract to manufacture bottles for John Clarke's Congress Spring. As demand for Saratoga's mineral spring water increased and fuel supplies near Syracuse became scarce, the brothers decided to move the business to Saratoga County to be closer to their best customer.

In a sense, Oscar Granger was returning home. He was born in the town of Moreau in 1803, but early in his childhood the Granger family moved to Vernon to open a general store. The Lake Desolation area was chosen as the new factory site due to its proximity to Saratoga, the availability of land and an ample supply of wood. The Grangers also found someone to help finance the move in nearby Jamesville. Entrepreneur John James had established two of the area's earliest paper mills ten years before and built a small village to support them. He also founded one of the county's first banks, which financed the Grangers' efforts to relocate their glass factory.

Due to the poor road conditions in the 1800s, small communities would very often be built close to large mills and factories, particularly those that were located far from an established city or village. Accordingly, when Oscar Granger first arrived at the site of the future glassworks, he immediately laid out plans to establish a small village with room for about two hundred people, complete with about twenty houses, a hotel for unmarried workers, a store, a gristmill, a school and a post office. The sawmill was built first on the southeastern side of the village to provide finished lumber for constructing the village. Later it was used to handle the vast quantities of wood used to fire the furnaces. A blacksmith shop and clay mill provided the tools and moldings needed in the factory. A state commissioners' survey in 1846 listed the Granger property as 140 acres of cultivated land and almost 1,000 of timber, valued at eight dollars per acre.

Within two years, the settlement of Mount Pleasant was completed and the factory commenced operations. To facilitate the slow and sometimes hazardous drive to Saratoga, an eleven-mile-long plank road was constructed from the village east to Porter's Corners, through Greenfield Center and on to Saratoga using three million board feet of lumber. To help pay down the cost of its construction, Granger installed four tollbooths along the road and charged travelers three cents at each tollgate. A shortcut down the mountain to Lake Desolation, called the "Dug Way," was cut through the thick forest that eventually led to the Kayaderosseras Creek in Middle Grove. Loaded wagons were lowered down this path with winches and rope. Once at the bottom, horses would be hitched to the wagons for the rest of the trip to Saratoga.

Local citizens were reluctant to become employed at the Saratoga Glass Works because most jobs required great skill and intense physical labor, so Granger sought glassblowers from England, Ireland and France. An average of forty-five employees worked seventy-two hours per week earning thirty-five dollars a month. French-Canadian lumberjacks supplied the four thousand cords of wood that were used each year. Most of the bottles were sold to the mineral spring companies in Saratoga, particularly the Congress Spring. The bottles were colored olive green and made to withstand the pressure of the carbonic gas contained in the mineral water. Lighter bottles of different colors were produced for ink, medicine, snuff, bitters and wine. During the 1850s, more than one hundred crates of bottles were brought to Saratoga each day (some sources quote a figure of 7.2 million bottles per year, but this seems unlikely, as it would equate to almost 20,000 bottles per day).

Life on "Glass Factory Mountain" was not easy for the workers or their families. The highest paid men were the five or six glassblowers, who worked long hours and sometimes had to eat five meals a day just to keep up their strength. Boys had to apprentice for many years at low pay in order to become

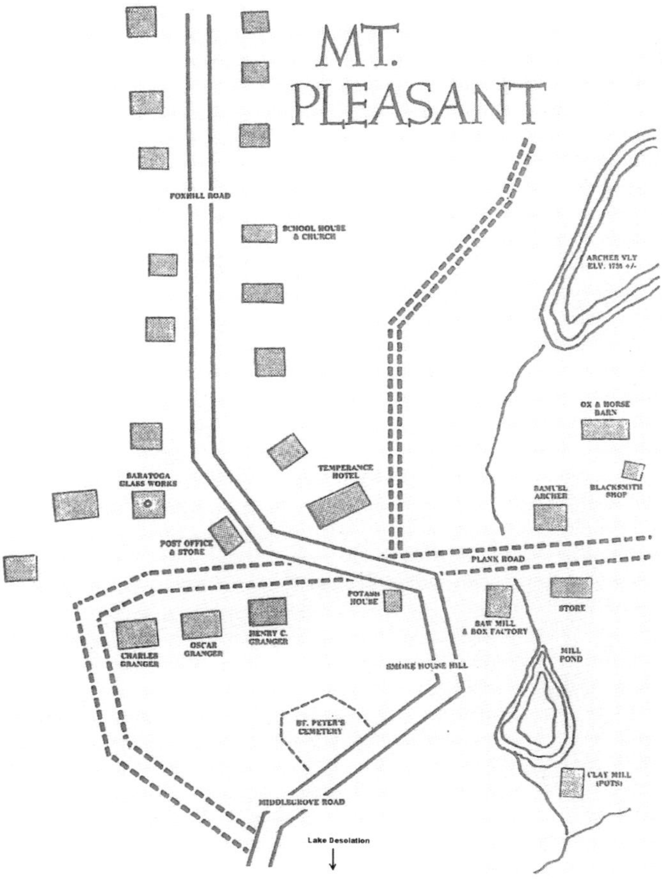

Map of Mount Pleasant soon after it was constructed. Very little evidence of this settlement survives today. *Courtesy of the New York State Archives.*

master glassmakers, often earning only fifty dollars per year plus room and board. To make matters worse, Oscar Granger was a deeply religious man who did not allow his employees to drink alcohol, gamble or socialize with single women. A sign posted at the factory proclaimed the "Rules & Regulations" that had to be followed. Winters were especially wearisome, since after the first snowfall there was generally no travel to nearby villages. Some children were sent to stay with relatives in Middle Grove for the winter so that they could attend school.

The little village of Mount Pleasant changed forever soon after the Civil War ended. A Rock City Falls paper manufacturer named Chauncey Kilmer purchased a controlling interest in both the Congress and Empire Springs in 1865. He then merged them into one stock company called the Congress-Empire Spring Company with a capital of $1 million. This consolidation coincided with the end of Granger's twenty-five-year contract to supply bottles. Kilmer was anxious to cut costs and told Granger that the contract would not be renewed unless he moved the glass factory to Saratoga to save on transportation expenses. Knowing that the factory could not survive the loss of its biggest customer, plans were made to establish a new factory within the village limits on the corner of Congress Avenue and Empire Street. It has been reported that several houses were physically moved from Mount Pleasant to a Saratoga suburb called Congressville, but this could not be independently verified. Nevertheless, Mount Pleasant became nearly extinct overnight, with only a few families remaining behind.

After the move to Saratoga, the product line was expanded to include pocket flasks, wine bottles and other containers. Sand for the mineral spring bottles was exported from the West at $100 per train car load; however, sand for the wine and other amber-colored bottles was found locally. Due to new production techniques, the employment force had been reduced to twenty-eight men and boys, including eight blowers and two melters, while production had increased to almost five thousand bottles per day.

The new glass factory operated for several more decades and was one of the stops for tourists who traveled to Saratoga during the summer gaming and racing season. One guide wrote,

> *If one has time, it is well worth while to look in upon the swarthy workers dipping their long iron tubes into the white heat of the furnace and drawing out the visceral mass, that the dexterous toil and distended cheeks they blow into good pints and quarts—there is a strange fascination about glass-making and as this establishment stands in the open fields, one may visit it without meeting the gloom and dusty heat of the city works.*

As the years passed, the factory became more and more obsolete compared to those that used automatic bottle-making machines. The Congress and Empire Spring Company held on for two more decades but was finally sold at public auction for $78,000. The City of Saratoga Springs purchased both the factory and Congress Park in 1911, giving the factory a brief reprieve from closure, but increased competition caused the glass factory to be shut down for good, ending seventy years of glassmaking in Saratoga County.

FROM SOAP TO NUTS

Although the towns of Saratoga Springs and Milton generated the greatest total economic output in most censuses throughout the 1800s, the town and village of Waterford arguably possessed the greatest variety of industry in the county. Its proximity to the industrial center of Troy and location at the vortex of several canals and rivers made it a perfect place for many manufacturing enterprises to flourish.

Very soon after the Champlain, Erie and King's Canals were completed, manufacturers arrived to take advantage of the new water powers. During the 1830s, a time when little other major industry besides farming existed in the county, the citizens of Waterford watched as more than a dozen factories were erected along the canals. These included a cotton mill, ink factory, flouring mill, soap factory, sawmill, machine shop, iron foundry, twine factory and fire engine shop. As soon as one enterprise shut down, another would quickly take its place. The machine shop was replaced by a strawboard factory; the twine factory was replaced by a nut factory; and the cotton mill was replaced by the King Dye Works.

The rapid replacement of manufacturers indicates that, although the water power of the canals was considered to be valuable and desirable, many of the businesses were small and did not last long. This also makes it difficult to form a complete history of Waterford's industries in any detail. For example, a building on King's Canal began its life as a button factory and then was converted in succession to a barley mill, feed mill and strawboard mill. The Mohawk and Hudson Paper Mill building was used as a foundry, axe factory, woolen mill and knitting mill. The long list of knitting mills sometimes refers to the same building under different owners and trade names, which has led to gaps in the village's industrial record. For example, little is known of the cement mill of Canvass White, even though he supplied his patented cement mixture to New York State for the construction of the Erie Canal.

Other businesses operated for decades but had several owners, which adds to the confusion. The Cable Flax Mill was established in 1812 and manufactured twine. It was founded by John Gue, who formed a partnership with Lucius Fletcher in 1831. When Gue retired, the firm became known as Fletcher and Parker. In 1862, the mill was purchased by Moses Hovey, and a short time later Edwin Hartshorn was admitted as a partner. Although the Cable Flax Mill is not well known, it was a prosperous concern for many years, manufacturing 1.2 million rounds of twine annually and employing 250 men.

The Waterford Sawing Mills of William Burton was one of the more interesting sawmills in the county. While most provided finished lumber supplied by Adirondack forests, Burton used such wood as mahogany and cedar from Mexico, rosewood from Brazil and satin wood from San Domingo. Cigar boxes were manufactured from Spanish cedar, while cabinets were made from Florida red cedar. In fact, nearly all of the lumber sold by the sawmill was imported from distant states and other countries. It was first established in 1835 by Hawes and Goodwin of Cohoes. Burton and his sons William and Benjamin purchased the machinery and goodwill in 1840 and moved the business to Waterford.

Perhaps the two most famous businesses in the history of Waterford were based on the inventions of their owners—the Button Fire Engine Company

Early photograph of the Button Fire Engine Works. *Courtesy of the Waterford Museum and Cultural Education Center.*

and the Eddy Valve Company. They were unique because there were few fire engine and valve manufacturers in the country, yet they succeeded for decades in Waterford. Only those that could offer cutting-edge innovations were able to survive in each of those highly specialized industries.

Lysander Button developed dozens of fire engine innovations at his factory on the King's Canal, although he was not the first to introduce the industry to the Capital Region. John Rogers began the manufacture of hand fire engines in Albany, relocating to Waterford around 1825. Button found a job there as a machinist, gradually moving up the ranks to become foreman, partner and finally sole owner. Most of his innovations were for the so-called "hand-pumper" fire engines. It is obvious that he possessed remarkable gifts in mechanical engineering, since by 1838 he had developed the first "piano engine," which delivered water from the front end of the trunk of the device, the "runaround" that returned water to the suction to relieve pressure on the hose, and the first fire engine water pump. In 1857, Robert Blake joined the firm and helped Button patent a variety of new inventions for steam fire engines, which were just making their debut. The partners' first few patents featured several improvements in hand-pump engines, but an 1862 patent improved upon steam engines by rearranging and increasing the diameter of the cylinders and modifying the air chamber.

Button's company flourished throughout the 1860s and 1870s, with deliveries made to all parts of the United States, Canada, South America, Europe and to local villages and towns. One study counted a total of one thousand steam fire engines in the United States, seven hundred of which were manufactured by Button. Six different sizes of fire engines were made, ranging from four thousand to six thousand pounds, featuring upright boilers designed to withstand high pressures, copper tubing and all of Button's many improvements. After spending almost half a century manufacturing fire engines, Lysander decided to retire and sell the entire works, including the rights to his many patents, to Holroyd and Company of Waterford. Years later, Button fire engines continued to be manufactured by the American Fire Engine Company, which was formed by the consolidation of four large manufacturers.

Another gifted inventor and businessman named George Eddy started the Mohawk and Hudson Iron Foundry and Machine Shop in 1847. At first, the company concentrated on car wheels and iron castings, but as time went on, other items were added to the product listing, such as wine and straw presses, chilled railroad wheels, lathe chucks, axes, hitching posts and cast-iron statuary. The foundry produced a stunning quantity of castings, including six thousand stoves per year.

Eddy patented a taper-seat, double-gate valve that would go on to become one of the most successful inventions in the county. Valves are used in all pipe lines to control the flow of water. The Eddy Valve consisted of two gates and a piece called the ball. The gates were hung on two trunnions that allowed them to revolve freely. This allowed the gates to seat in a different position every time to prevent them from becoming worn out in one spot. Eddy also perfected a parallel-seat, double-gate valve, a straightway valve and a check valve that kept water from flowing backward in case of a pipe break. An advertisement in the 1880s claimed that 90 percent of the towns in the country owned an Eddy valve.

As successful as his valve was, Eddy's hydrant was even more famous. The earliest model contained an important feature—an operating rod threaded below the ground line to prevent flooding in case the standpipe broke. The one-piece design was later modified to a two-piece design, and the main valve was changed to close against the water pressure, unlike other hydrant valves. Another key feature was the ability to remove the drip rod and drip valve through the top of the hydrant using ordinary tools for quick removal of obstructions, all without removing the main valve or shutting off the water from the mainline. Later, Eddy added the Bailey Fire Hydrant to the product list, featuring "brass mounting, a leather valve, and positive drip." Upon his retirement, a local businessman purchased the foundry and renamed it the Eddy Valve Company. For more than sixty years afterward, it remained locally owned and operated before being purchased by the Clow Valve Company of Oskaloosa, Iowa, which continues to manufacture the Eddy Hydrant.

Two of Waterford's most famous products—a fire engine from the Button Fire Engine Works and the Mohawk Hydrant from the Eddy Valve Company. *Courtesy of the author.*

Besides the Mohawk and Hudson Foundry of George Eddy, other foundries included the Gage Machine Works, J.M. King and Company, Holroyd and Company and the Globe Ironworks, manufacturers of steam boilers, bleachers' tanks and sheet iron. George Gage established the Gage Machine Works on the King's Canal in 1834, as well as an extensive building complex that at various times hosted the Brooks Nut Factory, Holroyd and Company, the Button Fire Engine Company and the King Dye Works. Twenty-five men were employed by Gage making lathes, machinists' tools, steam regulators, knitting machinery and paper machines. The foundry of J.M. King was established in 1829 and eventually employed eighty men for the manufacture of stocks, dies, taps, reamers and screw cutting machines. The founder, Daniel King, was a brother of John King of King's Canal fame.

It seems as though every industry imaginable was represented in some way within the town of Waterford, even if only by one factory. There were extensive gristmills, such as the Rock Island Flouring Mill of J.B. Enos and Company. Established in 1847, it employed twenty men and manufactured about four hundred barrels of flour per day and shipped them on its two canalboats direct to New York City. The flax and twine industry employed several hundred people until its use in packaging goods was replaced by paper bags and tape. Edwards and Younglove operated a strawboard factory with twelve men. The Mohawk and Hudson Paper Mill was established by Frank Gilbert on the King's Canal in 1872 to manufacture printing paper. Gilbert employed forty men at the mill and produced three tons of paper per day. Next door was the Brooks Nut Factory, which was established in 1835 and operated for over fifty years. Joshua and Elisha Morse established the Waterford Soap and Candle Factory in 1830 and became the largest soap factory north of Albany, with an output of ten thousand boxes of soap and candles per year. Other enterprises included a hay press factory, two lamp black makers, a sash and blind factory, a cement sewer pipe works and a veneer sawmill. Many of these small factories gradually went out of business at the turn of the century but were replaced by dozens of textile mills that eventually employed far more people.

The Country's Only Linen Thread Mill

On January 31, 1881, the real estate, mills, stores, tenement houses and water power of the American Linen Thread Company were sold at auction, bringing to a close yet another of Mechanicville's pioneer industries. The city had been the first in the world to host a linen thread factory, as well as the first in the country to manufacture striking matches and an improved form of roofing tarpaper. With the dissolution of the linen thread company, the last of the pioneer industries disappeared into history.

The American Linen Thread Company was first established in Lansingburgh in 1848, but the business languished with few customers. Three years later, it was moved to Cohoes and received new machinery imported from England. There the unique twines and threads were more fully developed and attracted the attention of several capitalists who invested a total of $30,000 into the plant.

For unknown reasons, the company relocated for the last time to Mechanicville the following year. Another $70,000 was invested by the partnership of William Wight, Abiram Fellows, Cramer Vernam, Philander Bennett and Job Viall. All of the partners except England-born Wight were Mechanicville natives. The partnership dissolved soon after, but the company continued to operate with little interruption despite the ownership changes. Lewis Smith joined as superintendent and remained for over twenty years. Before paper and clothing mills were established, the manufacture of linen thread was the only industry listed in Mechanicville that exported goods to other parts of the country.

The thread mill site was originally used as a woolen factory by Colonel Barrett and Squire Sutton of Albany, but it was not a success and burned down. Lowell Bullen and David Fairbanks purchased the property and erected a cotton mill. James Bradshaw and Company, which included Bradshaw, Fairbanks and Job Viall, owned the property until 1852.

Linen thread had limited but important uses for such things as fishing lines, linen cloth for lightweight clothing and the binding that connected the upper and lower portions of shoes. Since there were no other linen thread makers in the country, the mill flourished at its Mechanicville location, enjoying good management and favorable market conditions. Within a short time, the factory was employing eighty people and producing three hundred pounds of thread per day. The workforce was about equally split between men and women, although wages were unequal—about twenty dollars per month for men and nine dollars per month for women.

The company seems to have been well known in the business community of the country and received much attention for its patent product. A review of the new thread material was detailed in the November 1852 edition of *Valley Farmer* magazine:

> *Upon close examination of the various specimens of thread presented by this company, it is evident that skill directed their works. The extension and twisting of the filaments was performed with a uniformity surprising to the unpracticed eye. They seemed to be in parallelism of fibre for strength, held together by a firm twist, and possibly cemented together by the gluten of the plant. The samples of white thread showed a fine polished surface of silvery hue, very attractive—the bleached white was no less beautiful. The colored samples exhibit about 200 shades, many of them rich tints, though a few were dull—these colors were also highly finished. Judging the whole products of the American Linen Thread Company by the samples presented they are entitled to strong recommendation.*

The company received even more attention when its products were displayed at the state fair held in St. Louis. The thread was described as "among the most interesting objects exhibited at the New York State Fair—the article was pronounced by judges to be superior in strength and finish to any imported." It was also the only company of its kind to display its products at the Philadelphia Exposition, where American industry celebrated one hundred years of nationhood.

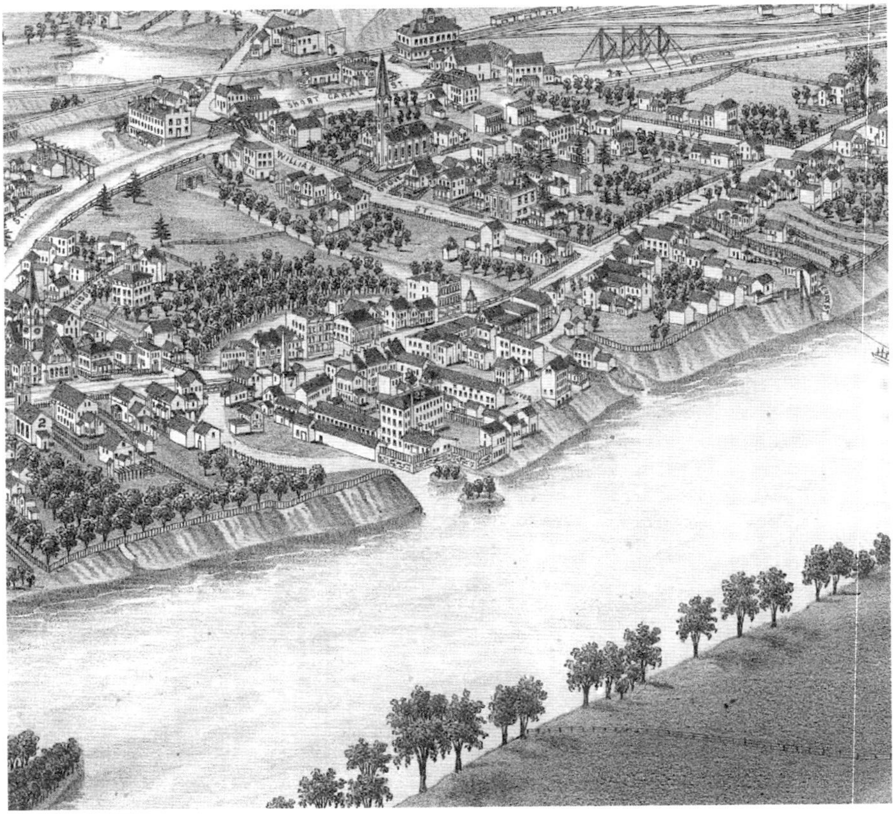

The American Linen Thread Company buildings were situated on the Hudson River (center, labeled "25"). The Champlain Canal is depicted at the top of the drawing. *Courtesy of the Library of Congress.*

Although the company was the only one in the country to manufacture linen thread, it was not a financial success for very long. The only period of prosperity came during the Civil War, when readymade thread was in high demand. The business was hampered by a severe drop in American flax production, the material used to manufacture thread. Until 1845, the domestic cultivation of flax amounted to 300 million pounds annually in New York State alone. In the year that the company moved to Mechanicville, the output had dropped to 940,000 pounds. Therefore, the flax used to make thread had to be imported from Ireland, Belgium and Holland.

The lack of high profits did not stop the owners from expanding the business over the years, however. An old gristmill first built by Rice Hulin of Round Lake was converted into a plaster mill. The company also owned a preparing mill, a gristmill, a sawmill and forty-five tenements for its workers. The original mill purchased by Viall and partners for $26,000 was expanded into a ninety-three-foot- by twenty-two-foot-long, three-story brick building. Another production building stood on the twelve-acre lot west of the canal that measured one hundred by forty-four feet in size. This building was located in a section of the village known as "Hell's Half Acre" from the notoriety of the taverns and canal workers who lived there. The three dams on the Anthony Kill generated ninety horsepower.

By 1880, the company had lost its domestic monopoly in linen thread, and debts began to accumulate rapidly. Two of its shareholders, William Van Schoonhoven of Troy and John Ross of Boston, were forced to file a foreclosure suit in the amount of $44,000. An auction was held in which the company was sold in two lots. The first lot included the main factory, warehouses, sawmill, gristmill, water wheel and machinery, and the second consisted of twelve acres of land that contained a three-story brick factory, two houses, a storehouse and additional water power rights. Van Schoonhoven and Ross filed the winning bids of $14,000 for lot one and $8,000 for lot two. It was a sad end to one of the county's most famous industries.

Healing the World

Saratoga's fame as a healing center with rare mineral springs, baths and an invigorating climate naturally had an air of healthy living. Nestled firmly into this environment was the G.F. Harvey Company, manufacturer of medicines that were shipped around the country. The long-lived business was highly respected and valued in the community not only for the many jobs that were created but also for its classification as a "light industry" that did not generate the unseemly smoke and wastes of other industries in Saratoga County. By 1890, the company was producing two million pills a day and 150 gallons of fluid extract, shipped to seven thousand physicians. Branch offices, used as distribution centers, were operating in Chattanooga, Chicago, Sioux City, Seattle and Los Angeles, among others. An additional manufacturing plant was established in Ontario, Canada.

George Harvey was born and raised in Vermont. In 1870, he moved to Wisconsin with several extended family members and worked for five years in the drug mercantile business. During that time, he observed the inequities and weaknesses in the drug industry, with individuals paying high prices for medicines that were spoiled or had little actual healing value. After several years of planning, Harvey decided to move to New York and establish a drug manufacturing and distribution company that would supply quality products at the lowest possible cost. He came to Saratoga Springs and set up a small business out of the kitchen of his Church Street home. As a one-man operation, he not only manufactured his products but also personally made sales trips to Vermont, Massachusetts and Upstate New York, traveling by horse and buggy. He later added several salesmen to go on the road for him as the manufacturing capacity of the company increased. He improved or invented machinery for compounding pills that automated many of the production processes formerly done by hand.

In 1889, Harvey formed a partnership with Sidney Rickard, a wealthy Saratoga businessman, and State Senator Edgar Brackett. It was incorporated under the name G.F. Harvey Company and capitalized at $250,000. A three-story brick building was constructed for a laboratory on Waterbury Street near the Adirondack Railroad, and rooms were set up in a building at 466 Broadway for offices and sales. Within five years, the product line had expanded to five hundred kinds of pills, granules, tablet triturates, fluid extracts, powders, suppositories, tinctures, syrups and elixirs. The company also carried a line of surgical instruments and physicians' supplies. New types of pill coatings were developed to maintain the drugs' virtues, and new packaging was developed for reliable distribution around the country.

The factory was a wonder of efficiency, cleanliness and automation, especially in relation to the general practices of the time. Almost two hundred people were kept employed as factory workers, bookkeepers, stenographers and salesmen. As one contemporary newspaper article put it, "George F.

The G.F. Harvey building on Waterbury Street, Saratoga Springs. *Courtesy of New York State Archives.*

Harvey and company have established a successful business because they are making something useful for which there is a demand by labor saving methods which make low prices possible; because their products are of the highest quality and purity with no deviation from exactness; because they have maintained the purpose of honest dealing in the finest sense of the term." One of their most well-known products was "Saratoga Ointment," marketed around the country.

The G.F. Harvey Company was one of the longest-lived and most respected businesses in the city's history. By the mid-1900s, the company had increased its capitalization to $850,000 and employed 150 factory workers and 65 traveling representatives. Since it was a local company employing local residents, a large portion of its revenues were distributed directly or indirectly throughout Saratoga Springs in the form of payroll and supplies needed to operate.

There were several other medicine manufacturers in the county, but none appears to have lasted beyond the life of its owner. One that lasted more than most was the medicine business of Beecher Hollow's John Latcher. He and several members of his family emigrated from Switzerland and first settled in Fulton County, New York, before moving to Edinburgh. After Arad Copeland died in 1884, Latcher took over his washboard factory and used it as a machine shop.

Latcher was a prolific inventor with more than a dozen patents to his name. The 1870 census listed him as a "Curry-Comb Manufacturer," which he patented two years earlier. Displaying remarkable versatility, he established a side business manufacturing medicines under the name J.W. Latcher and Company. One of his primary products was "Latcher's Anti-Dyspeptic Blood and Liver Pills," taken two to four times per day, "preferably in the evenings on going to bed, or until the bowels are very thoroughly evacuated." They were recommended for such ills as liver complaint, dyspepsia, blood and skin diseases, kidney problems, rheumatism, languor, dizziness and drowsiness. Another advertised product was "Latcher's Vitalizing Sarsaparilla," which could be taken to heal "general debility of the system, rheumatism, habitual costiveness, and chronic diseases generally." Some of his other many medicines included Mandrake (for jaundice, dysentery, dropsy and gonorrhea), Hops (to relieve pain and induce sleep), Burdock (for venereal, rheumatic and leprous diseases), Bitter Root (for chronic liver infections), Blue Flag (for liver and kidney infections) and Balmony (to remove worms).

A few other much smaller "patent medicine" enterprises existed in the county, but these were short-lived and are only known through entries in business directories. A "Great Tonic Laxative" called the "Wonder Laxatif" was manufactured by M.E. Raymond, Inc. of Ballston Spa and sold for twenty-five cents a bottle. The Lee Products Company of Saratoga Springs offered Lee's Liver and Kidney Tea for thirty cents per package to alleviate stomach and kidney troubles, as well as Lee's Ointment, "an antiseptic remedy of marvelous healing powers." *Lee's Guide to Saratoga* mentions the drugstore of T.H. Pennington, who manufactured Pennington's Norwegian Troches, "a most valuable remedy for diseases peculiar to the bronchial passages." A short biography of businessman Byron L. Hurd of Schuylerville claimed that in 1883 he patented his celebrated "Chemically Pure Baking Powder" and "Hurd's Adirondack White Pine Cough Balsam," both of which "he can hardly manufacture in quantities sufficient to meet the demand that exists for them in the United States." Unfortunately, there is no mention of these medicines outside of Schuylerville or at the patent office.

An unusual medicinal business flourished in Halfmoon for a few years. Reinshold Werner immigrated to America from Germany and, after operating a cooperage business for a few years in New York City, settled down in Halfmoon. There he commenced manufacturing malt liquor in a small distillery. His sons Herold and Henry joined the brewery when they came of age. In 1888, Herold established a side business producing what he called "malt medicines." There are indications that the malt medicine business enjoyed some popularity around the country at this time, as advertisements can be seen for malt tonics of every description.

Around the turn of the century, new laws were enacted that barred companies from making claims that could not be substantiated. Although these laws have not entirely stopped the miraculous healing claims of these so-called medicines even to this day, most of the businesses disappeared from view, leaving only respected and proven medicine manufacturers such as the G.F. Harvey Company.

The Paper Bag King

The paper mills of Milton were small compared to the mills found in Mechanicville and Corinth, but the combined output of over a dozen mills exceeded any other town in New York State for much of the 1800s. Much of the credit for this achievement is given to an English immigrant named George West, who acquired a group of mills along the Kayaderosseras Creek and became known as the "Paper Bag King."

Early in the century, nearly all paper was made by hand. The solution to the challenge of making paper from cheap raw materials was finally solved in 1840 by the introduction of the ground-wood process of pulp making, used by several local pulp mills, and ten years later the first use of a chemical pulp process, to be used by Ballston Spa's Glen Mill.

The nature of the pulp determines the type of paper that is produced. For example, the use of straw created a suitable paper for newspapers, such as was produced by Kilmer's Mill of Rock City Falls. Cotton rags were used to produce paper collars and cuffs, which required more strength and durability. Both the Eagle Mill and the Glen Paper Collar Company used cotton rags for the production of collars until they went out of style. Most of the paper mills of George West used leaves from the plantain family to create manila paper for the production of paper bags. Through new advances in technology, many modern mills now use waste to recycle paper and other products, reducing the demand for live trees.

Once the techniques of making paper from wood and straw had advanced enough to become profitable and mills could gather their own raw materials (chiefly by cutting down trees), paper mills in Saratoga County began to flourish.

Several paper mills were built along the Kayaderosseras while the industry was still young. One of the first paper mills in Saratoga County was established in Middle Grove by John James in 1836. Chauncey Kilmer built the second paper mill a few years later in Rock City Falls. He then built another mill a short distance to the east that became known as Kilmer's Mill; it was only the second in the country to produce paper from straw. Lindley Crane purchased a mill above Factory Village in 1845 that would become known as Eagle Mill, while Cady Hollister built the National Mill a short distance downstream.

The man who would one day own the largest paper bag operation in the world came to the United States from England almost penniless in 1849 at the age of twenty-six. His only advantages were skills learned from the papermaking trade in his home country and a determination to succeed in his adopted country. He first found employment in a New Jersey paper mill earning seven dollars per week. After one year, he moved to Tyringham, Massachusetts, and began work at a writing paper mill, where he invented machinery for cutting a watermark into the center of a sheet of paper. He took full charge of the J.R. Smith and Company mill of Russell, Massachusetts, in 1853 and then became part owner of a mill in Cummington. He later sold out his interests in order to seek out an area with greater opportunities.

In 1860, he brought his family to the town of Milton. For a short time, he worked at the Pioneer Paper Mill, one of the most successful in the county at the time. The following year, his employer, Coe Buchanan, offered him a chance to purchase the idle Empire Mill in nearby Rock City Falls. He decided to lease the property for one year to see if he could turn a profit. This he did and then purchased the mill outright. This was the start of a remarkable career that would span four decades.

West made a considerable impression upon the growth and prosperity of Ballston Spa and several hamlets near the Kayaderosseras Creek, mainly due to the demand generated from his line of paper bags. During this crucial time in forming the path of his success, he decided to manufacture high-quality manila paper that brought him customers from around the region. Most paper mills at the time used cotton rags as their raw material. When the Civil War and its aftermath created a scarcity of the commodity, many cotton mills closed down. West's decision to manufacture paper bags made of wood products using the latest technology was fortuitous and profitable. His design was superior to any on the market, and demand very quickly outstripped his ability to manufacture them.

The bags at his first paper bag factory on the corner of Milton Avenue and South Street were all made by hand in relatively small quantities. To

George West commenced building his "paper bag empire" in 1862 with the purchase of the Empire Mill in Rock City Falls. *Courtesy of the Chris Morley Collection.*

expand operations, West built a new bag factory in Rock City Falls adjacent to his Empire Mill. According to popular history, a man appeared one day at West's office in Rock City Falls and told him that he had a better way of making paper bags using machines. West gave him a job, and within a few weeks the machines were completed and running. He built Excelsior Mill next door to the Empire, as well as a mansion across the street. He then purchased and rebuilt the burned down Pioneer Mill and two mills in Middle Grove and renovated the Eagle Mill in Craneville.

In 1875, West made his largest purchase in Ballston Spa, bringing all of Jonas Hovey's holdings and mills under his management. He moved his bag factory into the brick building adjacent to the Union Mill, retrofitted the Glen Woolen Mill as a wood pulp mill and directed all of the paper mills upstream to manufacture manila paper to feed his new bag factory. West's "paper bag empire" then consisted of eight paper mills, a chemical sulphite mill and a paper bag factory. He also purchased and rebuilt a large paper mill in the town of Hadley that became one of the largest producers of manila paper in the country.

The success of his endeavors was dramatic. In the year that West took over his first mill, he had ten employees, a monthly payroll of $300 and shipped about a half ton of paper per day. Thirty years later, he employed over four hundred people with a monthly payroll of $14,000 and shipped forty tons of paper and 2.5 million paper bags per day. For many years he was the single largest employer in Ballston Spa and was well known throughout the region, especially during his years in public office as state assemblyman and congressman. On a trip to his native England, local newspapers noted that he was the largest paper bag manufacturer in the world and dubbed him the "Paper Bag King."

There is some confusion as to whether George West was the original inventor of the square-bottomed paper bag or some aspect of its production, such as the machinery. It is unfortunate that he did not think it necessary to apply for any patents during his career, which would have gone a long way toward determining his accomplishments. The honor of being the "official" inventor of the paper bag goes to Margaret Knight of Massachusetts, who patented a flat-bottom bag machine in 1871. However, there is ample evidence and firsthand accounts in the mid-1860s that indicate West was producing a type of paper bag at that time. Some sources also claim that West's nephew, James West, patented a paper bag machine and assigned it to him. For example, a *Saratogian* article in 1935 stated that "James West, who died recently, was the inventor of a machine to make the first square bottom paper bags. For fourteen years the Ballston Spa plant had a monopoly on the manufacture of these types of bags." This patent could not be found, however, and James was too young in the 1860s to have been involved with the industry. He did submit a few bag-related patents and assigned them to the Union Paper Bag Machine Company of Philadelphia. This company was formed in 1869 for the sole purpose of buying and fighting paper bag–related patents. Since George West was one of the founding members of this firm, it is likely that he indirectly "owned" the earliest patents in this field. Many years after he passed away, this fact may have morphed into the belief that West invented a type of paper bag, when in reality he simply had access to the first and most effective designs.

The consolidations that occurred at the turn of the century in various industries included the few paper bag manufacturing mills that were left. Only two of the original eight paper bag companies that operated in 1869

The Glen Mill in Ballston Spa was one of the county's few chemical pulp mills that used sulphite to break down wood into paper pulp. *Courtesy of the Chris Morley Collection.*

still existed in 1899. In the latter year, the Union Bag and Paper Company was formed in New Jersey with a capital of $27 million and absorbed most of the paper bag manufactories in the country. West received $1.5 million for all of his mills and officially retired. Since he was in his late seventies and had suffered from failing health for several years, this consolidation was a stroke of good timing for West and his family.

It turned out to be well timed in other ways as well. Since competition was increasing by the year and most of the profits to be made by small, one-man operations had already been realized, selling the mills at that time turned out to be a lucky decision. Almost immediately after acquiring West's mills, Union Bag and Paper began selling off those of Middle Grove and Rock City Falls. By 1912, only the four paper mills and bag factory in Ballston Spa were still owned by the corporation. Worker strikes after World War I and increased transportation costs sealed the

Interior view of the Island Mill in Ballston Spa, owned by George West. *Courtesy of the Chris Morley Collection.*

fate of these last mills, and the corporate officers put them all up for sale and gradually moved its bag-making operations to Hudson Falls and Fenimore. Although some of the mills would open up again under new management, they never again achieved the success and stability provided by George West.

The World's Largest Book Paper Mill

The origins of the paper industry in Mechanicville can be traced back to a huge dam project undertaken in 1882, led by Horace Medbery of the Glen Paper Collar Company. He and several other investors decided that a location on the Hudson River near that village would generate enough power to operate a large paper mill that they planned to erect on the site and, in addition, have other manufactories purchase the excess power. The Hudson River Water Power and Paper Company commenced building the massive dam and paper mill complex using bricks that were manufactured on site.

Despite a recession and flood that damaged the structure, the dam was completed the following year. By that time, some of the early investors had departed, but a Scotland native named Thomas Duncan joined the firm and became its treasurer and superintendent. The mill had until that time produced chemical pulp fiber in an unfinished form. The product would then be sold to other mills that had the ability to convert the pulp into paper. One of Duncan's goals was to expand the mill so that it could complete both processes. His efforts were so effective that the awkward name of the enterprise was changed to the Duncan Company in 1895.

After years of expansion, the mill was the largest employer in Mechanicville, with about five hundred men on the payroll. Four machines were added to take advantage of the new chemical sulphite process for making paper, and production increased to almost forty tons of book paper per day. Lumber was brought in by boat from company woodlands in the Adirondacks, and finished paper was transported out on the Delaware and Hudson Railroad. Sales offices were opened in New York City and Hartford, Connecticut.

The Duncan Company brought unprecedented prosperity to Mechanicville, causing the village to double in population over a fifteen-year period. The enterprise had many advantages: a first-class power source, a seemingly inexhaustible supply of raw materials close at hand, an active and experienced management team and connections to the outside world by canal, river and railroad. By 1900, it was shipping 1.5 million tons of paper and 800,000 tons of sulphite pulp annually, with revenues increasing almost every year. Duncan's son John joined the firm, bringing with him valuable insights from his education and training in Europe. The company was also able to generate income by selling power to several businesses located near the dam.

Changes were taking place within the paper industry, however, that threatened the stability of mills throughout Saratoga County. Huge paper conglomerates were formed in order to compete in a worldwide market, causing many small mills to be consolidated into larger entities. Since the officers and directors of these and other corporations were located in far-off cities with little understanding or concern for the local economies of the individual mills, it was more likely that decisions would be made for the sole benefit of the stockholders, sometimes having disastrous effects.

Fortunately for Mechanicville, this would not be the case for many years to come. Its merger with the West Virginia Pulp and Paper Company in 1904 would actually benefit the community, with some exceptions. The new company (known shorthand as Westvaco) was founded by Scotland native William Luke as the Piedmont Paper Company. After several acquisitions and aggressive management, the company was incorporated as Westvaco in 1899 and continued to seek other expansion opportunities. It purchased the Duncan mill for $3 million and immediately announced plans to modernize and enlarge the facility. Some controversy arose when most of the Duncan managers were dismissed a year later, but the expanded facilities provided employment for hundreds of others. An equipment inventory listed thirty-five 1,600-pound beating machines; eight refining engines; five Fourdrinier machines; and five different-sized super calenders, capable of producing ninety tons of paper every twenty-four hours.

A nationwide economic downturn in 1907 led to the next bump in the road for the mill. When production slowed down, workers were immediately dismissed to maintain profits. Additionally, while the former Duncan Company had few documented labor disputes or production disruptions, managers for Westvaco were not as prone to bargain with employees. Until the turn of the century, most factories operated on the two-tour system, in which shifts were alternated weekly between eleven hours and thirteen hours

Construction work at the West Virginia Pulp and Paper Company. *Courtesy of the Stillwater Town Historian.*

to remain in production twenty-four hours a day. When other mills began transitioning to a three-shift system and Westvaco management refused to follow suit, the calendering department at the mill went on strike. The strikers were quickly fired. Other departments went on strike in support but were gradually replaced by "scabs" and workers brought in from other cities. As a result of the strike, many families left Mechanicville forever, while others remained but commuted to work in Schenectady, Waterford or Troy.

In succeeding years, conditions began to improve. Production expanded in the years before and after World War I until it was the largest of the six mills in the Westvaco empire. It was also the largest in New York State, and some newspaper accounts reported that it was the largest book paper mill in the world for a time. By 1920, the capacity had increased to 120 tons of book paper, 75 tons of soda fiber and 75 tons of sulphite fiber daily. The mill employed about eight hundred blue-collar and white-collar workers and generated millions of dollars for the local economy.

The Great Depression hit the paper industry hard, as people did not have the discretionary funds to purchase books or newspapers. However, Westvaco was able to avoid bankruptcy, unlike many other paper mills in Saratoga County, and even remained profitable throughout the 1930s. Its contracts with several large publishers such as Hearst and Crowell-Collier enabled it to maintain a steady production.

The end of the Depression and the economic expansion that followed World War II led to a corresponding expansion of the Mechanicville mill, with another three hundred workers added to the payroll and additional funds poured into new machinery. The huge Delaware and Hudson rail yard located nearby, a well-trained workforce and continued access to inexpensive lumber combined to ensure that the mill complex would remain in operation for decades.

The prosperity of the mill and the city of Mechanicville peaked during the 1950s. Employment held steady at almost 1,400 workers who generated about $15 million in revenues. The mill produced high-grade white paper that was used for envelopes, copying and even postage stamps. A new process called the Neutral Sulphite Semi-Chemical method enabled management to use many other types of wood in papermaking that expanded the supply of available lumber.

Several issues arose during this time that mitigated the prosperity and stability provided by Westvaco. The union movement gained in strength so that the Mechanicville mill finally became unionized. However, District 50 of the United Mine Workers Union had little experience negotiating on behalf of papermakers. This led to inferior employment contracts relative to other paper mills, even others within the Westvaco organization. Wages were sometimes 10 to 12 percent less than at other paper mills in the Upstate New York region.

Two long and painful strikes prompted workers to switch to the United Pulp and Paperworkers Union, which called for yet another strike. These strikes were reminiscent of the bitter International Paper mill strikes of Corinth in the 1920s that included multiple court orders, the use of "scabs," sabotage and violence. But when they were over, both sides came to the conclusion that strikes were a measure of last resort, and from that point forward more emphasis was placed on negotiation and compromise.

Operations continued much as before, although by the late 1960s so much of the plant had become automated that the workforce was down to about seven hundred. Most local residents believed the mill would continue to operate indefinitely, but it turned out that its closure was imminent for a variety of reasons. Unlike their parents, many young people in the area had no interest in a career with long hours and hard manual labor. The Adirondack forests were slowly becoming depleted of easily accessible lumber, and new restrictions were made by the state to limit the discharge of pollution into the Hudson River. The mill itself was over one hundred years old and would need significant capital investments to remain competitive.

In 1971, Westvaco abruptly announced that it was closing the Mechanicville operation. A few efforts were made in the years that followed to reestablish the plant, but a recession, high tax assessments and the factors listed above presented obstacles that any large-scale manufacturer could not overcome. It was another heavy blow to the once-vaunted paper industry of Saratoga County.

The Country's Largest Newsprint Mill

A common theme in the development of industry throughout Saratoga County was the location of a suitable water supply for power and transportation. The Hudson River in the Palmer Falls–Corinth area had plenty of potential to generate the power needed to operate any business enterprise, especially once a dam was constructed that could regulate the seasonal flows of the river. Ira Haskins utilized the Hudson to power his sawmill as early as 1804, and in the years to follow several grist- and woolen mills were erected nearby. In 1829, Beriah Palmer purchased seven hundred acres of land on both sides of the river adjacent to the main headwater and created the Palmer Water Power Company. His goal was to seek investors who would build mills along this tract of land. However, the only taker was Thomas Brown, a trustee of the company who operated a woolen mill and hard edge tool factory for some years before being accidentally shot by one of his night watchmen.

Although the Palmer Water Power Company failed to attract the large number of investors and entrepreneurs that had been envisioned, the promotional activities of that concern and the construction of the Adirondack Railroad garnered the attention of the German-born brothers Alberto, Albrecht and Rudolph Pagenstecher, who had entered the United States in 1866 bearing the revolutionary Keller-Voelter wood pulp grinder. The brothers first utilized this machine in Curtisville, Massachusetts, but a lack of wood and an insufficient water supply in the Berkshires prompted them to look elsewhere to expand their operations. In 1869, the Pagenstechers and several investors leased land in Palmer Falls and established the Hudson River Pulp Company. This gave the brothers access to one of the best

The Brown Woolen Factory of Corinth a year before being purchased by the Hudson River Pulp and Paper Company. *Courtesy of the Corinth Town Historian.*

potential water powers in the state, as well as millions of acres of forests in the nearby Adirondacks.

By the name they gave to their new company, it is evident that the brothers originally did not plan to produce paper in Palmer. Historically, pulp and paper had always been produced in separate facilities. However, within a year of commencing operations, a sixty-eight-inch-wide paper machine for producing writing paper was installed in the former Thomas Brown tool factory, so the Hudson mill may have been the first in the country to manufacture both pulp and paper at the same site. For this reason, as well as the use of the Keller-Voelter grinder and other innovations, the creation of the Hudson River Pulp Company is widely considered to mark the beginning of the modern paper industry.

The full potential of the water power at the site took a number of years. At first, the mill used a wooden raceway leading from the Hudson that produced about seven hundred horsepower. The simple structure was washed away by a flood in 1877 and rebuilt in stone. A few years later, a 150-foot-long crib dam was built that increased the fall to 87 feet and significantly increased the power that could be generated for machinery, which in turn permitted the rapid expansion of the paper mill. Another flood inflicted heavy damage, leading to the construction of a new concrete dam that generated twenty-five thousand horsepower, enough to reliably power most mill operations well into the twentieth century.

An experienced paper mill manager named Warren Curtis moved to Corinth in 1871 and became superintendent. Curtis spearheaded the expansion and ongoing modernization of the mill, often inventing new machinery and techniques that increased production. A short time after his arrival, he added another mill building to house two additional paper machines. By 1880, the mill complex was employing over two hundred men and producing fifty tons of paper and seventy tons of wood pulp per week. Curtis also decided to switch production from writing paper to the more profitable line of newsprint paper and added the word "paper" to the name of the company.

Expansion continued almost unabated during the following decade. A brick office building was constructed to accommodate the growing administrative requirements of running such a large mill, while additional wood grinders and paper machines were added to the factory. Five of the new sulphite pulp digesters, which used chemicals to break down wood into paper pulp, were installed, as well as a 112-inch-wide machine that was considered to be one of the largest in the world. Enough finished product was being generated to justify the construction of a rail spur from Corinth Station to the mill property.

Although the Hudson River often produced plenty of power, the huge machines then in operation needed a tremendous rush of water to remain in operation. There were times when some of the wood grinders had to be shut down during low water in the summer months. In order to maintain operations, the mill began to overproduce pulp in the spring, which was converted into sheets and stored dry in huge piles. During a dry spell, water could be added to these sheets and mixed in beating machines to create usable pulp for the paper machines.

In 1896, two new paper machines were installed in the factory, known as No. 1 and No. 2. These were the most modern machines available and remained in operation for decades, which increased the production of newsprint paper even further. Two years later, the Hudson River Pulp and Paper Company and sixteen other mills were purchased by the International Paper Company during the wave of "trusts" that swept the nation. At the time of its formation, International Paper became the largest paper company in the world, and the Hudson River plant was its largest subsidiary. Other area mills that joined International Paper included the Glens Falls Paper Company and the Lake George Paper Company of Ticonderoga.

A major remodeling project was undertaken in 1902 to install several one-thousand-watt General Electric generators. An equipment inventory listed nineteen beating machines, twelve refining engines and seven Fourdrinier machines ranging in size from 63 inches to 147 inches, capable of producing 135 tons of paper every twenty-four hours. The mill's primary customers for its newsprint paper were the *Philadelphia Enquirer*, the *Sunday Mirror* and the *New York Herald Tribune*.

The mill was forced to purchase pulp sheets from other paper mills to meet the demand by the new machines. To solve the problem, Warren Curtis and his son established a ground wood mill at Curtis Falls, a quarter of a mile upstream from Palmer Falls. The Curtis Manufacturing Company obtained a contract with International Paper to produce all of its ground wood pulp requirements. A new rail spur was constructed off the Delaware and Hudson mainline to the Curtis mill.

During the early part of the 1900s, the mill was the leading producer of both wood pulp and newsprint in the country and was by far the single most important factor in the economy of Corinth and the surrounding area. More than half of the citizens of the village either worked in the mill or were directly supported by one of its employees. Approximately 60 percent of village and school taxes were paid by International Paper. The two giant piles of reserve pulp wood, measuring over one thousand feet long and towering eight stories high, were visible from many parts of Corinth and became part of the village's landscape, a testament to the productive capacity of the mill.

The organization of labor unions in many industries coincided with the trust movement of the late 1800s. The Corinth mill was among the first of International Paper's holdings to identify with labor groups, which included Palmer Local No. 7 of the International Brotherhood of Paper-Makers (IBPM) and Local No. 4 of the International Brotherhood of Pulp, Sulphite and Paper Mill Workers (IBPSPMW). One of the most influential union managers was Jeremiah Carey, who served as president of the IBPM from 1905 to 1923.

Throughout the 1910s, labor and management largely worked together in harmony, and there were few shutdowns or disturbances. This prosperous and cooperative era came to an end as the country entered a recession stemming from the end of World War I. Unions had grown in strength during the war as millions of men went overseas to fight, leaving acute labor shortages back home. When the war ended and soldiers returned to work, union leaders

Massive quantities of wood were needed to keep the Hudson River Pulp and Paper Company supplied with raw materials. *Courtesy of the Corinth Town Historian.*

were slow to realize that their bargaining power was curtailed. In addition, the tariffs that protected domestic paper from foreign competition was lifted in 1913, so that by the early 1920s Canadian imports amounted to over 1.1 million tons, up from just 25,000 tons a decade earlier.

In 1921, the two unions submitted a contract that requested a 10 percent increase in wages. Facing a drop in demand for paper, mill owners reacted with a 30 percent pay cut and the elimination of overtime pay. IBPM and IBPSPMW membership voted to reject the proposal and go on strike, and some twenty-five thousand workers walked out, including those of the Hudson River mill.

The situation in Corinth degraded rapidly when International Paper announced that it would restart the mill using nonunion workers. The initial attempt at restarting failed when no workers dared to show up for work. However, management persisted and brought workers in from as far away as Michigan and provided housing for them on mill property. International Paper attempted to lure workers back with bonuses and instituted lawsuits against various village agencies that were supporting the strikers. In the meantime, the mill began to switch production from newsprint paper to specialty papers used for magazines and catalogues as the paper industry experienced yet another transition.

Despite its best efforts, International Paper was unable to lure back very many striking workers and posted large financial losses. However, as the months turned into years and management showed no signs of relenting, the strike gradually lost community support and was finally called off in 1926. The bitterness instilled in the striking workers, some of whom never returned to the mill, would last for years afterward.

Once mill operations returned to normal, new machinery was installed to support the conversion to coated paper. Two new, 145-inch-wide super calendars were purchased, while the paper machines already in operation were gradually converted. An off-machine coating process was used to spray a layer of coating on sheets of paper as they passed by and then dried by moving the paper around a large wheel. Much of the finished product was used for labeling bottles and cans, but it was also used for comic book covers, shelf paper and wallpaper base. A plant for bleaching sulphite pulp was built in 1938, while the production of paper cores (used to roll paper for shipment) amounted to eight million feet per year.

In the years following World War II, the machines were again upgraded. Machine No. 2 was given an on-machine roll coater, while the speed of machines Nos. 3 and 4 were increased from 800 to 1,250 feet per minute. A part of the mill that was constructed in 1888 was taken down and replaced with a modern, 500-foot-long building that housed a single machine. Known as Machine No. 11, it was capable of producing a 208-inch-wide sheet at a speed of 2,300 feet per minute. It would be the last new machine purchased for the mill.

The output of the Hudson River mill during the mid-1900s was staggering. Because it was the only International Paper–owned mill that manufactured coated papers, it had to supply all of the orders received by the parent company. In the mid-1960s, the mill was turning out 175,000 tons of paper annually with a workforce approaching 1,700 employees.

Yet another modernization program was instituted in the 1970s, although no new machines were added. As the century drew to a close, profits began to drop as the cost of maintaining such a large complex of aging buildings began to weigh on the finances. The age of water-powered mills had drawn to a close long before, and even though International Paper had kept its flagship mill as modern as possible, its location and construction were no longer feasible. By 2001, the number of employees had dwindled to just two hundred, while the one remaining paper machine in operation was forty-four years old. The mill was shut down for good the following year, bringing an end to over one hundred years of papermaking in Corinth.

MINING AND QUARRYING

Geological activity over the past millions of years has blessed Saratoga County with a vast mineral wealth, especially in the northern towns of Greenfield and Hadley. The rocks found at or near the surface in the county are ancient, which has resulted in rare fossil sites and an abundance of minerals. While these minerals have not been found in large concentrations, they have tempted many individuals to invest in mining and quarrying operations for almost a century.

One of the earliest attempts to establish a mine in the county was done by a man named Dr. Mitchell from Northampton, Fulton County. Around 1840, he built a separator on Mount Antonio close to where the Sacandaga and Hudson Rivers unite. Examinations were made at various places on the mountain that convinced the doctor that valuable deposits of iron could be profitably extracted, but his failing health led him to abandon the effort. A few years later, the firm Bushnell and Palmer of New York City opened a mine on the southeast side of the mountain and built a blast furnace at Hadley about one mile away. Although the iron was of superior quality, the expense of hauling it to the canal in Glens Falls was too high to make the operation profitable.

In 1880, P.H. Cowen of Essex County commenced operations about one hundred feet below the Bushnell and Palmer site. Before long, he hit a vein of iron ore that looked very promising. He sent about one hundred tons of rock to the Burden Iron Works testing facility in Troy, which reported that the sample yielded 51 percent "Bessemer quality pig-iron." The construction of the Adirondack Railroad in the years since the first attempts at mining the mountain provided inexpensive transportation, so Cowen calculated that he could mine the ore at a cost of $1.90 per ton and sell it for $5.00 per ton. As with most of the mines in Saratoga County, however, the calculations were too optimistic, and profits were not forthcoming.

Another feldspar quarry was operated two miles north of Batchellerville by the Claspka Mining Company of Trenton, New Jersey. It opened in 1906 and contained a large outcropping of granite pegmatite, which was excavated in two pits. The large, lower pit was excavated to a depth of fifteen feet. It was reported that quartz was discovered in "pure masses several feet across," while feldspar was "light-gray microline, found in pure masses four feet across." Other minerals found in large quantities were muscovite, biotite and beryl.

The American Feldspar and Mining Company attempted to establish a feldspar mine in the town of Corinth on the Adirondack Railroad in 1899. Feldspar is a common raw material (it is the most common group of minerals found in the Earth's crust) used for producing ceramics and glass. The operation consisted of a small, lower pit and an upper, larger one on the slope of a hill. It appears that the mine never got very far past the experimental stages, however.

In 1906, the Glens Falls Graphite Company opened a mine in the town of Day, about one mile west of Conklingville. It was reached by an eight-mile drive from Hadley on the Delaware and Hudson Railroad. A mill was built 1,700 feet to the southwest of the mine and connected to it by a wagon road. A small dam was constructed in the nearby stream so that the mine could be partially powered by water. Some of the equipment listed included a Sturdevant crusher, Sturdevant rolls, wet screens, hexagonal revolving screens, a steam-powered dryer and equipment for bolting and grading. The mine itself consisted of a single open pit on the slope of a hill. At the time of a United States Geological Survey, the excavations had only reached about 20 feet underground and covered about an acre of land. The mine and mill were taken over by the Sacandaga Graphite Company, but the site was not as fruitful as initial tests indicated and was abandoned a few years later.

The mines of Greenfield never produced the profits that many were seeking, but operations lasted for almost a quarter century. Around 1900, William Carver was prospecting near Porter's Corners and came across some black lead. He related his discovery to a businessman named Meyer from New York City, who formed a stock company with the purpose of mining the lead. When graphite was also discovered on the property, Meyer formed the Empire Graphite Company and switched over to mining for graphite. Excited local and New York City financiers invested $500,000 in the company to build an extensive mill complex, purchase equipment and hire workers. The heavy mining equipment was delivered to North Greenfield by railroad and had to be manually hauled the remaining several miles with teams of horses. Delivering the eleven-ton drier was such a spectacle that it was recorded for posterity on several postcards.

The Porter's Corners mine covered about two acres of land with a maximum depth of twenty-five feet. At least one tunnel was excavated three hundred feet into the ground. Rocks were brought out of the mine using small railcars, which delivered them to a gyratory crusher and two sets of crushing rolls. The pulverized rock was then moved to "buddles," or troughs used to wash and separate minerals, and later to settling tanks, rotary sills, wet screens, rotary driers and sizing screens in succession. Since the mill was situated on a steep slope of a hill, gravity was used to transport the treated graphite to the horse teams that hauled it to King's Station four miles away.

It took some time to achieve full production at the mine. Seven years after Carver first made his discovery, the Empire Mill was still in the experimental stages. Initial plans for the mine were quite ambitious—machinery was installed with the expectation of netting one hundred tons of graphite per day with fifty men. Although these numbers were never reached, the

The Empire Graphite Company mine just outside of Porter's Corners in Greenfield. *Courtesy of the Milton Town Historian Collection.*

mine was deemed a success when the graphite was sold for $1 per pound. A fire destroyed the original mill in 1909, causing $100,000 in losses, but a larger one was built in its place. Unfortunately, soon after the new mill recommenced operations, prices dropped and forced it into bankruptcy.

Other entrepreneurs must have thought there was still money to be made, because the mine was purchased by the Flake Graphite Company in 1917, and operations resumed once more. The risks involved with any type of mining are illustrated by the company's experiences, as related by the *New York State Museum Bulletin.* These included the discovery of a horizontal "drift" of graphite, where it was planned to "block out the ore between this and the surface, but as the turn from the direction of the dip to that of the strike was made too near the portal, there was little ore to be had." The bulletin also stated that "further east, in the loose ground, a drift was attempted, but was poorly planned and will probably be abandoned. On the south side of the knoll, there are three abandoned holes." Market prices did not help the situation. During World War I, the price had increased to sixty cents per pound. However, once the war ended and other mines were discovered around the country, the price gradually fell to only three cents per pound. It was estimated that 2.5 million tons of graphite were available for mining when the Flake Graphite Company took over the mill, but the plunge in prices prevented the company from achieving much success. Perhaps someday, when other mineral deposits around the world become depleted, the mining industry in Saratoga County will be revived.

Foundries and Fabricators

The Capital Region's largest foundries were located in Troy, where a large steel-making industry blossomed in the mid-1800s. A combination of abundant ore supplies from the Adirondacks, plentiful molding sand from the Mohawk Valley and Troy's location on the Hudson River made it one of the leading iron- and steel-producing centers. The country's first steel using the Bessemer process was made in Troy, while just down the river Henry Burden was manufacturing millions of horseshoes with his patented machines. Troy and Albany together manufactured hundreds of thousands of heating and cooking stoves that were sold around the world. One of the Troy foundries even fabricated the armor plating for the first ironclad warship, USS *Monitor*.

Since the village of Waterford is located just a few short miles from Troy, it also had these advantages. Therefore, some of the largest and most diversified foundries in Saratoga County operated in Waterford for over a century. Perhaps the most famous was the Mohawk and Hudson Foundry established by George Eddy in 1847. At first the company produced car wheels, stoves and iron castings of his own designs. As time went on and the manufacturing process was refined, other items were added to the product listing, such as wine and straw presses, chilled railroad wheels, lathe chucks, axes, hitching posts and cast-iron statuary. In 1873, Eddy patented his world-famous taper-seat, double-gate valve (called the Eddy Valve), and he later patented an improved fire hydrant (called the Mohawk Hydrant). These proved so successful that they soon accounted for most of the foundry's output. Eddy employed up to two hundred men for nearly thirty years.

The stock and die manufacturing company of the Holroyd family operated for more than half a century. James and Mary Ann Holroyd emigrated from England to Waterford in 1831. Upon his arrival, James immediately commenced the manufacture of blacksmiths' stocks and dies and later organized the firm Platt and Holroyd. After receiving an education in the public school system, his son William joined the business to learn the trade of die-maker. After only a few years, the business was manufacturing all types of blacksmith dies, such as taper and plug taps, axle stock dies, combination axle cutters, pipe taps and reamers, solid die plates and bottoming taps. These tools were well known for quality and shipped all over the United States and many foreign countries. The firm employed forty to fifty men and generated approximately $70,000 in revenue per year. The Holroyds purchased the Button Fire Engine Company in 1882 and operated it for about ten years before selling it off to a national corporation.

The leaders and residents of Saratoga Springs have been sensitive to the introduction of heavy industry within city limits that may negatively impact the tourist trade, so it is somewhat surprising that one of the county's largest foundries was located there. Because the Baker and Shevlin foundry was moved to about one mile away from the shopping district, however, its presence does not seem to have been an issue. The foundry was first established on Putnam Street when business partners James Baker and George Shevlin purchased the Putnam Street Foundry in 1881. Orders for machinery were received at a steady rate until, by 1905, the company had outgrown its facilities. Plans were made to purchase twenty-eight acres of land just south of the city along present-day Route 50 and the mainline of the Delaware and Hudson Railroad. A sizeable powerhouse was constructed and equipped with a 150-horsepower Corliss engine and a 100-kilowatt generator. Other buildings included a 350-foot-long machine shop, a 200-foot-long foundry and a pattern storage house measuring 100 by 50 feet. The two-story office contained fireproof vaults and a large drafting room, and all of the buildings were made of brick and iron with slate roofs to be as fireproof as possible.

While machines and parts of all kinds were manufactured, machinery used in paper and pulp mills was the foundry's specialty. One example was reported in the *Watertown Daily Times*, which stated that the Baker and Shevlin foundry was the sole manufacturer of a new combined chipper and crusher invented by Howard Shortt of Carthage, New York. The machine was being adopted by many sulphite pulp mills from northern New York State to West Virginia. The foundry also manufactured machines for the

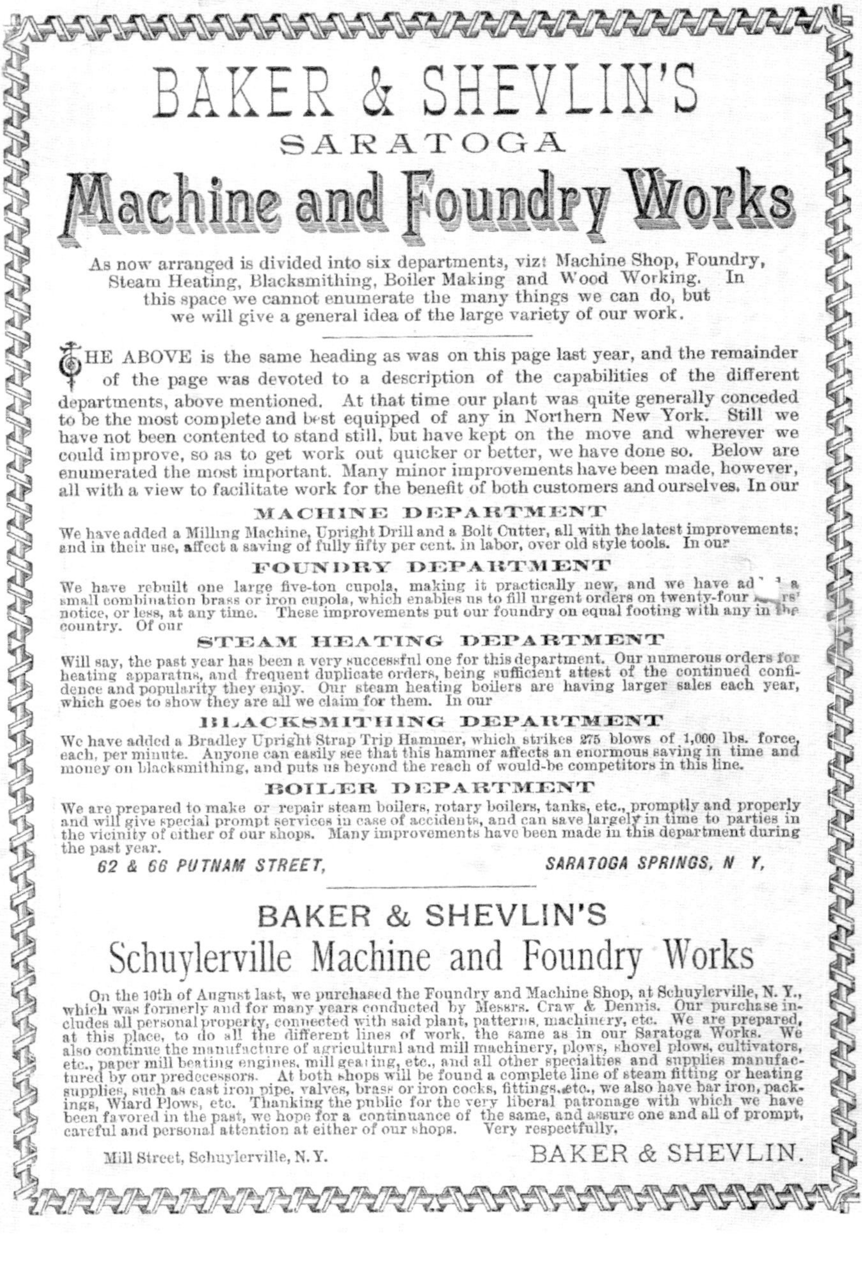

Advertisement for the Baker and Shevlin "Machine and Foundry Works" from the 1890 *Saratogian* Almanac. *Courtesy of the author.*

J.M. Voith Company, originally a German concern that was taken over by the United States government under the Alien Property Custodian Act.

The foundry's paper machine sales and service proved to be quite profitable, and when the partners decided to expand operations, the papermaking center of Schuylerville seemed to be a logical choice. An old foundry was located there, established by David French in 1832. He commenced operations in a former distillery building and in the basement of a woolen factory. A boy named David Craw worked at the foundry from the beginning and remained there through various partnership changes. When French retired, the business was turned over to Craw, and it became known as David Craw and Company. The foundry was known for operating over half a century without ever suspending operations. About ten men were employed in the fabrication of agricultural tools and paper mill machinery. When Craw retired to become a farmer, he sold the foundry to Baker and Shevlin, who went on to operate it for another twenty years.

James Baker retired from day-to-day operations in 1912 and sold his share of the business to Shevlin. For three years the company was operated under the name George F. Shevlin Company, but it was forced into bankruptcy. James's son Frederick, who had ample experience running the business with his father, then assumed control and renamed it the Baker Manufacturing Company. Through wise management, he was able to build the business back up to its former level. Frederick successfully operated the foundry for fifteen years until his death at the age of sixty-five. His father had retired due to poor health and moved to Daytona Beach, Florida. The plant sat idle in the years following Frederick Baker's death until it was purchased by the Espey Manufacturing and Electrical Corporation.

Another large foundry was located in nearby Ballston Spa and operated by the talented Uline brothers. It was established in a brick building on Bath Street in 1896 by Charles and Merrill Uline for making various finished goods from raw steel (prior to the Ulines, the foundry was owned by John Reynolds, the grandfather of *60 Minutes* commentator Andy Rooney). This was a successful business for many years due to the specialized knowledge of its founders. As an example of the plant's capabilities, the owners had several automobiles built almost from scratch for their own use. The brothers' most popular product was the Uline Chilled Plow, described by them as "the best implement of the kind that has been invented, which has worked an important revolution in the structure of plows."

In early 1905, the business was purchased by Henry Davison of Springfield, Massachusetts, and William Namack of Utica, New York. It then became the Davison and Namack Foundry Company, employing over one hundred workers. Business was so good that the new owners opened another foundry on Lewis Street, which would later become their primary plant, producing fifty tons of castings per day. Almost half of their work was for the General Electric Company in Schenectady. One notable success was producing a locomotive shield casting that other foundries, including those that specialized in locomotives, could not create due to its complexity.

The town of Galway had limited industry, but a successful foundry was operated on a small creek in Mosherville for many years. The members of the Mosher family were Quakers whose church was located a short distance east of Barkersville Road. In 1839, Reuben leased a half acre of land to his oldest son, Orin, who built a foundry and blacksmith shop. A large, nineteen-foot waterwheel was used to produce the equivalent of eight horsepower, although the low flow of the river in the dry season would sometimes be insufficient to turn the wheel.

Orin's younger brother Israel joined the foundry and soon after invented an improved shovel plow, which was used to loosen the soil between rows of crops for aeration and weed control. The plow soon became the primary product of the foundry. Orin Mosher died unexpectedly at the age of thirty-

The Mosher Plow Works was Galway's largest industry before being converted into a cheese factory. Today it survives as a general store. *Courtesy of the author.*

eight and left his estate to his wife, Mary, who sold her interest in the foundry to Israel for $500. Israel and another brother named Philip continued to manufacture Israel's shovel plow, as evidenced by an advertisement in a local newspaper called the *Home Register*. In the June 1856 edition, the brothers stated that Mosher's "lately improved plow is now being offered" and that "all other kinds of castings are made to order, on short notice." In the same paper was an advertisement for the general store that was situated adjacent to the foundry and run by another Mosher brother named Silas.

The foundry was sold by the Mosher family but continued to operate for many years under various owners. In 1870, it was owned by William Cornell and had four employees. The foundry manufactured about 750 plows and other castings annually until 1889, when it was sold to Calvin Whiting of Albuquerque, New Mexico. Frank Fonda assumed control of the property a few years later, running the foundry as well as the general store next door. Clarence Cornell, son of previous foundry owner William, remained as the molder throughout the ownership changes until it finally shut down in 1926 to become a country store.

The foundry industry in Saratoga County and the rest of the Capital Region employed tens of thousands of people throughout the 1800s and early 1900s. However, huge ore deposits were found in the western states where labor was cheaper, so much of the steel-making and fabricating operations in the Northeast gradually shut down or were relocated.

The Waterford Knitting Mills

For years the Victory Manufacturing Company was the only large cotton mill in the county. When cotton shortages stemming from the Civil War caused most of the cotton mills to shut down, the paper industry rose in importance for the following five decades. When the paper mills began to close down due to intense competition from larger, more efficient mills located in cities, the textile industry came full circle as the manufacture of clothing began to gain in importance. Citizens in Stillwater, Corinth, Mechanicville and Ballston Spa were gratified to see several large factories establish themselves in the production of such articles as shirts, gloves, socks and knit underwear.

The textile mills of the Cohoes-Troy-Waterford area numbered in the dozens and employed tens of thousands of people. Cohoes, which borders Waterford in the county of Albany, contained the world's largest cotton mill that alone employed five thousand people. Waterford obviously benefited not only from its canal water power but also from its proximity to Troy, the "Collar City," and Cohoes, the "Spindle City."

While a few clothing factories were established earlier, most began operations in Waterford near the turn of the century, long after extensive factories had been started up in nearby Cohoes and Troy. In 1880, an overview of the industry in the Waterford *Sentinel* listed the Diamond Knitting Mill of J.W. Himes, E.G. Munson's mill, the Waterford Knitting Mill, the Hudson Valley Knitting Mill and the "Burr Mill" of Luke Kavanaugh, so named because he invented an improved knitting burr. All of these enterprises were situated along the Champlain or King's Canal and were powered by water. This small list was greatly expanded beginning in the late 1800s.

While many Waterford textile factories were small or short-lived, many others were large and successful. The Waterford Knitting Company employed 200 people manufacturing fine, all-wool scarlet, natural and camel's hair underwear for men and women. It was incorporated in 1885 with a capital stock of $50,000 and managed for many years by Thomas Breslin. The factory contained 2,270 spindles and thirty-five cylinder knitting machines in a four-story brick building. The officers of Waterford Knitting were also the officers of the Ormsby Textile Company, manufacturers of men's and women's wool underwear beginning in 1893. The Pilot Knitting Mill, established in 1875 by Van Schoonhoven and Company, employed about 90 people. The Hudson Valley Knitting Company in the northern part of town on the Champlain Canal employed over 125 people and turned out 250,000 shirts, drawers and hosiery per year. The Massasoit Knitting Mill of Edward Munson ran six sets of machinery that produced over 300,000 men's and women's knit wear annually. It was founded in 1872, employed 100 people and operated for thirty years.

The Kavanaugh Knitting Company was one of the largest mills in Waterford's history. Both Luke Kavanaugh and his wife, Mary, were born in Ireland and set sail for the United States soon after their son Charles was born in 1854. Luke found a job as a machinist with the foundry of Gage and Campbell. Kavanaugh's time at the machine shop was evidently profitable, as he amassed a small fortune of $20,000 by 1870 (almost half a million dollars today). Much of his success was due to his patents, such as an improvement in blades for knitting machine sinkers, an improvement in burrs for knitting machines and a knitting burr blade with a distinctive shape.

Finally deciding to make use of his inventions at his own mill, Kavanaugh established the Bishopton Knitting Company in 1886. The manufacturing plant consisted of ten cylinder knitting machines and twenty-four web sections to produce fine men's and women's knit underwear. A machine shop was located at one end of the building to manufacture Kavanaugh's knitting burr, which was sold to mills around the country.

Kavanaugh and his son Charles greatly expanded the family's business enterprises by erecting a large, three-story brick building for the Kavanaugh Knitting Company. It became a significant manufacturer of knit underwear and employed hundreds of people. The mill was equipped with thirty-three knitting cylinders and sixty web sections. The chief product was fine balbriggan underwear, but cotton and wool underwear were also manufactured as demand warranted. For several years, the company supplied all of the cotton undershirts used by the United States Army and Navy. In 1908, another building was erected as a spinning plant to furnish the yarns that were used in the knitting factory. The additional building made it possible for the company to control all aspects of the clothing production cycle: spinning the raw cotton into thread, weaving the thread into cloth and finally sewing the cloth into garments. When Luke Kavanaugh passed away, the business passed to his sons, Charles and Frederick. Upon their retirement in 1920, it was sold to the Goodyear Rubber Company.

The Ford Manufacturing Company was established by John Ford in 1891 on Fourth Street. After the wooden building was completely destroyed by fire a few years later, the business was moved to North Second Street. In 1919, it was purchased by Robert Reis and Company of New York City. The factory thereafter conducted all of the knitting, bleaching, dyeing, cutting, finishing and boxing of the popular Reis Perma-Sized men's tee shirts, athletic shirts, sport shirts and briefs. More than 250 people were employed at the mill producing $250,000 worth of goods per year.

The former Laughlin Textile Mill on Mohawk Street was also used by the Waterford Knitting Company and the Ormsby-Morris Company. *Courtesy of the author.*

Throughout the late 1800s and early 1900s, Waterford was home to some fifty knitting mills. Admittedly, many of these were short-lived, small in size or both. Even a simple listing would be tedious, but a few of the largest are given here. The J.W. Himes Underwear Company was established in 1890 and employed about 500 people. Prior to that company, Himes was involved with the Diamond Knitting Mills, which employed 250, and the Waterford Knitting Company mentioned above. The Eureka Knitting Mills, located on Second Street, was started in 1881 by the partnership of Ford and Pynes and employed 350. The Novelty Knitting Company took over the former Button Fire Engine Company building on Third Street in 1894 and employed 100. The Hudson Valley Knitting Company also employed 100 people on Hudson Street. Just a few of the mills that have been all but lost to history are the Clover Knitting Company (Hudson Street), Clyde Knitting Company (Fourth and Division Streets), Capital Mills, Inc. (Second Street), Waterford Woolen Company (Ballston Street), James Ranney Dye Works (Fourth and Ballston Streets) and the Waterford Yarn Company (King's Canal). Sidney Sault established a factory to supply the textile mills with shipping boxes in 1892. His plant employed 20 men and produced over 100,000 boxes per month.

Although the clothing mills enjoyed great success for several decades, they ultimately could not compete with the low-cost labor markets of the southern states and overseas. Some held on late into the twentieth century but in the end could not retain workers very long in an industry that paid low wages and offered sporadic employment.

Socks and Gloves from Ballston Spa

As the Union Bag and Paper Company closed the last of George West's former paper mills in 1915, local residents wondered if the village would survive. The paper mills had provided hundreds of jobs for three generations of workers. The Ballston Scythe, Axe and Tool Works had been destroyed years earlier, while electric-powered factories located in large cities made it increasingly difficult for the water-powered industries of Ballston to compete. The village was facing the prospect of high unemployment and rapid depopulation.

Fortunately, two individuals came to Ballston Spa around this time and went on to establish large clothing factories. One was Floyd Shutts, who established the Ballston-Stillwater Knitting Company, manufacturers of knit socks (described below), and the other was a Gloversville businessman named Albert Ackerman. In 1918, Ackerman established the Ballston Knit Glove Company in a small building on South Street and began producing women's plain gloves and men's seamless knit gloves. With a combination of frugality and wise investing, the business quickly expanded into the Union Store property, the Sweet Furniture store and the old Temperance Hall. Ackerman then established branch facilities in Schuylerville and Oneonta, while many private residences throughout Ballston Spa served as miniature "branch locations," as the fingering work was done by employees from their own homes.

For several years the company continued to expand and flourish. A setback came in 1930, when Ackerman was struck by an illness that briefly left him incapacitated. This came at a time when the company was experiencing severe financial difficulties, and with Ackerman out of commission, there was no one with enough experience and authority to take charge of the situation. A receiver was finally appointed to manage the finances, and Ackerman's partners reorganized the business under the name Ackshand Knitting Company (derived from combining the three partners' names together).

The company survived, and Ackerman recovered from his illness to lead it to even greater success. He patented several improvements in manufacturing gloves, including one that was used to supply seamless knit gloves and knit caps to the United States Army and Navy during World War II. By that time, the business employed seven hundred men and women and had a monthly payroll of $50,000. When Ackerman retired in 1956, it was the second-largest knit glove maker in the country. A year later, the new directors attempted to merge with the Zwicker Knitting Mill of Wisconsin, but the action was blocked by antitrust laws. Frustrated, the owners sold the mill's equipment to the Gloversville Knitting Company, and Ackshand Knitting closed its doors for good.

Floyd Shutts of Amsterdam had a wholesale business called the F.J. Shutts Company, dealing in hosiery, underwear and sweaters. One day, he placed a large order for work socks, but the manufacturer did not have the capacity to meet his deadline. This event convinced Shutts that there was an opportunity for a clothing manufacturer in this type of market. Therefore, he established a small knitting company in 1918 to produce knitted wool work socks. James Beverley and Richard Currie from his former company joined him in the new venture.

The original factory was on the site of the old Artesian Lithia Spring on Saratoga Avenue. A wing fronting on Ralph Street was added to accommodate the growing business. Within three years, the mill occupied the entire block and contained 122,000 square feet of floor space. The first products were woolen mixed "half hose" manufactured at a rate of 1.2 million per year, later increased to 18 million pairs using modern machinery. The plant had its own siding off the Delaware and Hudson Railroad to receive raw materials and ship finished goods around the country. It was one of the largest employers at the time, keeping five hundred people at work. The New York firm of Clift and Goodrich was the exclusive selling agent for the domestic and foreign markets until 1926, when Shutts formed his own sales organization.

In 1929, the Ballston Knitting Company merged with the Stillwater Knitting Company to reduce costs and overhead. The assets of the Stillwater

The Ballston Knitting Mill complex on Ralph Street in the mid-1900s. *Courtesy of the Paul and Marilyn Pastore Collection.*

Knitting Company were sold at auction in order to raise funds for the preservation of the Ballston plant, which at the time was having financial difficulties. Although there was some doubt as to whether the company would survive, the crisis was short and the merger turned out to be successful.

Shutts remained president until his death in 1934. The fate of the business was briefly in limbo when a man named Archie Peisch purchased it and made plans to close down and move the machinery south. However, a friend of the late owner named Walter Mooney decided to buy the mill from Peisch and take over its operations personally. He became president, and the other two original partners remained behind to help him operate it—Richard Currie as vice-president and James Beverly as secretary and treasurer. Despite concerns over the transfer of management, the company seemed to be on a strong footing two years later. Seven full-time salesmen found markets for its products in all forty-eight states. Production in 1936 amounted to 37,590 dozen sweaters valued at $343,000 and 1,270,983 dozen socks valued at $1.83 million. The output of socks made up almost half of the entire country's production.

World War II interrupted the flow of raw materials for many industries around the country. The manufacture of sweaters was discontinued at both the Ballston and Stillwater facilities as the nation concentrated on a wartime production. Thereafter, the knit company focused on athletic and work socks. An office was established on Worth Street in New York City, later moved to the Empire State Building.

Employees of the mill were constantly faced with the threat that the company would move its operations to the southern states where labor was cheaper. As early as the 1920s, many northern-based textile companies were announcing plans to do so. However, the company continued to generate profits and even had some of its greatest success in the 1950s. There were a total of four main production buildings and eight warehouses. The company ran its own spinning plant (the Ballston Spa Spinning Company) that produced yarn of all types from approximately three million pounds of wool, cotton and mixed fiber products per year. The Ballston plant at that time employed 360 workers and shipped more than fifteen million socks.

The company was finally forced to follow the lead of other textile companies and move production south. The mill in Ballston Spa continued to operate in support of the business until 1994, having such customers as Kinney Shoes and K-Mart. However, even more intense competition from Asia caused the company to sell its property to Brown Wooten Mills. Less than a year later, Brown filed for Chapter 11 bankruptcy but then reorganized under the direction of James Mooney, who formed Saratoga Woolen Mills. The new company managed to achieve sales of $1.5 million. Ironically, one of Mooney's biggest problems was finding enough workers to meet the demand for his woolen socks. Wages were about $7 an hour, and the mill's factory environment was said to be loud, dirty and hot. Some applicants asked if they could be laid off in the spring so they could take higher-paying jobs in construction and landscaping. Production was therefore limited to what could be made from fourteen of the factory's twenty spinning frames. A combination of razor-thin profit margins, foreign competition and local labor shortages forced Mooney to close after only a few years of operation, and the decades-old textile industry in Ballston Spa came to an end.

Clothing a Nation

Although Waterford possessed the most extensive clothing industry in the county, several important clothing mills were established elsewhere in the county. Those of Saratoga Springs, Mechanicville and Corinth in particular operated throughout much of the twentieth century and gave employment to hundreds of people.

While Saratoga Springs has achieved much greater fame for its mineral springs and horse racing than its industry, the city was still home to several of the largest manufacturers in the county at one time. One of these was the city's first and only large clothing factory, known as the Van Raalte Mill on Excelsior Avenue. The mill was likely one of the few in the world that contained a mineral spring inside the building that was accessible to all of the employees.

As with several other enterprises in the city, Senator Edgar Brackett was instrumental in first bringing the business to Saratoga. Joseph Clark first established a business for making gloves made of silk fabric in the small town of Augusville, New York. After his factory burned down, he relocated to nearby Sidney. When the demand for his product exceeded the capacity of the building he occupied, Senator Brackett persuaded Clark to take advantage of the ample labor available in Saratoga (which had about fourteen thousand residents at the time) and construct a brand-new building a short distance from the center of the city. In 1905, the Empire Spring on Excelsior Avenue was put on the auction block, and Clark purchased it for $70,000. In order to fit in with the high-end architecture of the city, he built an attractive red and white lime brick building right over the spring and featured two of the stone lions that originally guarded the Congress Hotel. Local residents were also fond of Clark's cow, which grazed contentedly in the courtyard for many years.

The new Clark Textile Mill carried out the complete production cycle of each glove, including cutting the patterns, sewing, dyeing, finishing and boxing for shipment. The gloves were made from silk imported in bales from Japan and stored in a vault for safekeeping. Two strands of silk were twisted together and the yarn was boiled to remove the seracin, which was the natural preservative produced by the silkworm when spinning its cocoon. The resulting yarn was then unwound and baled for shipment to the United States.

The business was a success for Clark, only closing down briefly in 1917 due to a workers' strike. After almost fifteen years, Clark retired and sold the plant to Emanuel and Zealie Van Raalte of New York City for $1.8 million. It is rumored that Senator Brackett, who helped finance the Clark mill, used his profits from the sale of the business to build the McGregor Golf Course.

The Great Depression and slackening demand for silk in general caused the mill to close in 1927 and remain so for about four years. However, Ray Carmer, who acted as assistant superintendent for the Van Raaltes, returned to Saratoga to reopen the mill to produce women's mesh hosiery. Demand for this product eventually declined, but this time the owners moved quickly to find a different clothing line. A new material called nylon came into wide use, so the company refitted to begin manufacturing undergarments and, by 1936, was selling this product exclusively. At its peak, the company employed eight hundred people and generated more than $1 million in revenues annually using 135,000 square feet of floor space. A section of town on Excelsior Avenue became a desirable neighborhood in which most of the residents worked at the Van Raalte Mill for decades on end.

A new building on High Rock Avenue was in the process of being constructed in 1968 when the Van Raaltes sold the business to Cluett, Peabody and Company, a Troy-based clothing manufacturer. For a few years it seemed as though the mill was as strong as ever, but its fortunes began to change in the 1970s. The corporate directors began looking for a buyer as sales declined, and when it became apparent that the mill was in danger of closing for good, forty employees and a few investors banded together to assume ownership. For over ten years it was known as Saratoga Textile Inc., but in 1986 it was sold to North Carolina–based Falk Fiber and Fabrics, Inc. To the disappointment of employees, the new owners simply wanted to remove the machinery and equipment to sell at a profit elsewhere.

Knitting machines of the Van Raalte mill in Saratoga Springs. *Courtesy of the Brookside Museum, Saratoga County Historical Society.*

Mechanicville's textile industry was eclipsed in importance by the West Virginia Pulp and Paper plant but still employed hundreds of people in several medium-sized factories. The first of these was the shirt factory of C.F. Crosby located on First Street. Later it was moved to the Crosby Opera House and purchased by E.E. Ide and Company of Troy, which made shirts and collars. Pierce Kane erected a small knitting mill in 1865 that produced three hundred shirts and drawers per day. Eight years later, Ephraim Newland and William Denison established the Saratoga Knitting Mills (consisting of the Denison Mill and the Newland Mill) to manufacture women's hosiery and men's underwear. In 1889, the firm was reorganized under the name Newland, Denison and Company and concentrated on children's knit underwear. Almost one thousand articles were produced each day by 75 men and women. The mill expanded over the years so that by the time it burned down in 1904, it made $100,000 per year and employed over 150 people. Charles Crosby, who opened two shirt factories in the 1870s, was the pioneer clothing manufacturer of Mechanicville. In 1886, he was induced to relocate to nearby Washington County to take advantage of the lower taxes, but a few years later he returned to Mechanicville and set up a new shirt factory.

One of the longest-lived textile firms in Mechanicville's history (rivaled only by the Kay Dunhill Company mill) was the Mechanicville Knitting Company, organized by Edward Strang in 1892 for the manufacture of men's medium and heavy weight shirts and heavy woolen underwear. His partners included George Rogers, Edgar Holmes, Morris Myers and Joseph Packer. A large mill was built at the end of Viall Avenue near the railroad passenger station and the Commercial Hotel.

The timing of Strang's new venture appeared to be unfortunate. The following year witnessed a stock market crash that led to an economic depression. However, while other companies declared bankruptcy, Mechanicville Knitting prospered and grew at a slow but steady pace. Once the depression ended, the pace of growth increased so that by the outbreak of World War I it had doubled its market capitalization. During the war, the company shipped tens of thousands of underwear overseas to American soldiers fighting in the European Theater. Strang and three successive presidents led the company with conservative yet flexible management that allowed it to operate nearly continuously for over forty years. For example, when tastes in clothing changed after the war, the company dropped the underwear line and began producing children's sleeping garments. These were marketed mainly in New York City and known as Kozy Kids sleepwear.

At the death of Edward Strang in 1917, his son Arthur was elected president and served until his own death ten years later. Successive presidents were Courtland Rogers, Fred Clute and Osborne Baker. While other textile firms experienced frequent and damaging worker strikes and disputes, management at Mechanicville Knitting practiced a liberal compensation program that regularly shared profits with its employees, as well as paid out generous dividends to its shareholders. For example, an employee bonus program in 1914 awarded each worker with an additional week's pay. The next year, each worker received a 4 percent bonus.

By 1930, the company was producing about two thousand garments per day. The raw materials used each year included 250,000 pounds of cloth, 100 million yards of thread and 2.2 million buttons. Between eighty and ninety men and women were employed throughout the Depression of the 1930s. It was during these difficult years that many textile manufacturers were forced to move south to take advantage of the lower wages paid in that

Interior view of a shirt factory in Stillwater. *Courtesy of the Stillwater Town Historian.*

The Pitkin Shirt Factory operated for a short time before being purchased by Cluett, Peabody and Company of Troy. *Courtesy of the Corinth Town Historian.*

part of the country. Mechanicville Knitting was able to hold on for a few more years by marketing a new type of clothing called sweatshirts. However, despite the Strang family's best efforts, the corporation was finally dissolved in 1942, and the building itself was destroyed by fire in 1967. Only the Kay Dunhill dress factory continued to operate in the city.

The clothing industry of Corinth was never very extensive, but one mill in particular operated for over seventy-five years. The Pitkin brothers purchased a building on Mill Street in 1897 and attempted to operate a shirt factory. There was plenty of expertise in the new venture—it was started up under the direction of George Ide of the successful collar manufacturing house in Troy and managed by his nephew Nathan Ide. The company seemed to have a winning business plan, consisting of manufacturing shirts and selling them to the Ide Company factory in Troy. Despite these advantages, the mill shut down after only six or seven months.

Two other shirt companies were established on the site of the Ide factory soon after, but neither succeeded. In 1899, another Troy-based firm named Cluett, Peabody and Company purchased the building outright from the Pitkins and immediately expanded the structure and added more machinery. The belts and pulleys of the machines that ran the length of the main production room were all run by steam engines. Light was provided by heavy kerosene lamps suspended from the ceiling.

For a few years, the mill consisted of the downstairs machine room, several stitching rooms in which buttons were sewn on by hand and a small extension on the northeast corner for the boiler. By 1937, another six additions had been made to expand the stitching rooms and add offices, storage and an infirmary. The largest construction project was carried out in that year, consisting of a 103- by 65-foot addition made of steel and brick that according to Cluett officials made it "the most modern factory in Northern New York." It was equipped with seventeen drinking fountains and a recreation room "for the health and happiness" of the workers. Another 150 employees were added to the payroll once the addition was complete.

For the life of the mill, the cloth was cut in Troy and shipped to Corinth for final finishing. Raw stock arrived in Corinth by train and was taken by horse and wagon to the factory about a mile distant. Trucks started to be used in 1915, making one trip per week. During the early 1900s, production amounted to about 6,000 shirts per week, each week consisting of six ten-hour days. By mid-decade, the output had increased to 7,200 per day, with employees working five eight-hour days. The products consisted of fancy and white shirts, army shirts, gray navy shirts, twill and flannel.

In the 1940s, Cluett, Peabody and Company was the largest shirt maker in the country, with more than 4,000 total employees. The Corinth branch's 250 employees performed what was called "piecework." For example, some workers received twenty-one cents for every dozen pockets sewn onto the shirts. While many men and women were grateful for the work, conditions in the factory were said to be harsh. The machines were so loud that people had to yell to be heard. The building itself was hot in the summer; one employee brought a thermometer to work, and it registered 110 degrees. The company gave away free salt pills and water so that the employees would not get sick.

Succumbing to trends in the clothing industry, Cluett, Peabody shut the plant down in 1975, a year before a federal statute would have given workers a vested right in their pension fund. Some saw their pension payments drop 75 percent from what they expected. It was a heavy blow to former employees and to the community as a whole as another industry disappeared.

Bleaching on Peebles Island

An island at the south end of Waterford was used by Native Americans as a seasonal fishing and trading camp. The first owners were Goosen Van Schaick and Philip Schuyler, who purchased it in 1664 when it was part of the Halfmoon Patent. Earthwork fortifications were built on the island during the Revolutionary War, since it was the last line of defense on the military road along the Hudson River. In the early 1800s, a dam was constructed from Waterford village near First Street that supplied power for a flour mill. However, the dam became useless in 1823 when the Champlain Canal was put into operation, as it raised the level of the water over the dam itself. Another dam was built by John King for his King's Canal project that reached Peebles Island, but no other mills were built there after the loss of the flour mill. The Van Schaick family continued to hold on to it until 1909, when it was sold to Cluett, Peabody and Company.

The island purchase was just another expansion of the giant Troy-based shirt manufacturer, which at various times had branch locations in Saratoga Springs, Mechanicville and Corinth. The company required huge quantities of white cloth in its shirt and collar operations, so it was decided that building a bleaching plant would be less expensive in the long run than purchasing bleached cloth from other vendors.

Troy resident Sanford Cluett invented a process for creating shrink-proof cloth and stretchable paper that eventually received the trade name of "Sanforizing." This compressive shrinking technique was developed at the new bleachery. Demand for Cluett, Peabody's shrink-proof shirts caused sales to soar in the mid-1900s. The company claimed that it produced enough cloth in one year to form a piece that would have been one yard wide and nineteen thousand miles long, all of which came from the Peebles Island mill.

The bleachery was located in a two-story brick building with about sixty-five thousand square feet of floor space. To support the heavy machinery, the brick walls were constructed two feet thick at basement level, one and a half feet thick at ground level and one foot thick on the second story. The heaviest machinery was restricted to the basement and first floor, supported by a heavy timber structural system and strategically placed steel columns and plating. The second floor, which was used by the testing and inspecting departments, contained large windows measuring six feet wide by fourteen feet high along the entire length of the building to let in as much natural light as possible.

Perhaps the greatest strength of the bleachery was that it contained all of the manufacturing operations under one roof. Material was easily sent from one department to another in a smooth and continuous chain. The other structures of the plant (administration, storage, utilities, etc.) were offset from

The former Cluett, Peabody and Company bleachery on Peebles Island. *Courtesy of the author.*

the main building. Although the Sanforizing process had not been developed at the time the factory was built, it proved to be adaptable and was able to house the new machinery without many structural changes.

The processes used to properly bleach and prepare the cloth were complex. Baled cloth from weaving mills (most of which came from Cluett's own mill in North Grosvenor Dale, Connecticut) was received in the Grey Room to be unbundled and stitched together in continuous units averaging twenty-three miles in length. The units of cloth were then subjected to a precise gas flame process to remove loose fibers. The process had the added benefit of making the cloth more impervious to soiling later.

The next processes involved chemical treatments. After a thorough washing and de-starching, the cloth was treated with a strong solution of caustic soda to give it a permanent luster that kept its appearance even after many washings. The bleaching department then used hydrogen peroxide to give the cloth a pure white color. A continuous "J-Box" method of bleaching was used that boiled the cloth in caustic soda, washed it clean and boiled it in peroxide, all using one machine.

In the Finishing Department, the cloth was squeezed dry and put in drying cans to completely remove all liquids. Some cloth used for handkerchiefs was placed into framing and calendaring machines to straighten it with an electronic eye device and iron it smooth under intense heat.

The Compressive Shrinking Department on the second floor tested the Sanforized method to ensure that the cloth would not shrink once it left the factory. A sample wash test of each batch would determine exactly how much each would shrink to within a quarter-inch to the yard. Once it was proven that the cloth met the company's shrinkage standards, it was folded into webs and stamped with the identifying lot and style number. Each lot was carefully inspected to make sure there were no faults or soiling. If any faults were found, the lot was sent back to the Bleach House on the first floor. After the Examining Department certified each lot, the Shipping Department loaded them into open shipping trucks or cartons to either be stored or sent to other Cluett, Peabody factories to be made into shirts and collars.

Although the bleachery was considered to be small for its kind, it received high marks for producing quality material at low costs. However, by 1960, other, more modern bleacheries were able to deliver quality cloth less expensively than at the Peebles mill. The company's final improvement to the mill was an addition to the bleachery to handle the new "wash and wear" fabric that was becoming popular. Despite the addition, costs continued to rise relative to the competition, and Cluett, Peabody was forced to cease operations in 1972.

Even with the presence of a large factory using a variety of chemicals for sixty years, portions of Peebles Island lay relatively unspoiled. In 1973, it was added to the National Register of Historic Places due to the former presence of a Mohican Indian village and Revolutionary War–era earthworks. That same year, New York State purchased the entire 138-acre island using funds from the Environmental Quality Bond Act. Today it is known as the Peebles Island State Park and features views of the Hudson and Mohawk Rivers and the population centers of Lansingburgh and Waterford.

Burned Out

Fire Takes Its Toll

Until the advent and widespread use of automatic fire sprinklers, factories were extremely vulnerable to fire, and it was rare indeed for one to survive unscathed for more than a decade or two without experiencing some damage or total destruction. Nearly all mills were constructed of wood until the late 1800s, even the paper mills. Most also had stockpiles of lumber and other wood products for fuel or raw materials. Any number of events could result in a damaging fire, including arson, sparks from machinery, lightning, overturned lamps and spontaneous combustion from the interior heat of large piles of sawdust or wood chips. Even the most prepared mills were not immune.

One notable example was the Glens Falls Paper Mill, located on the south bank of the Hudson River in South Glens Falls. In an 1877 description of the plant, managers boasted that "as the steam is constantly kept up in the establishment, the means of extinguishing a fire, should one commence, are ample." Fire pumps and three hundred feet of hose were kept in position, while several more fire pumps were close at hand at the neighboring Morgan Lumber Company. However, the mill could not have been prepared for the event in July 1883 that destroyed the entire mill. A rotary boiler used for cooking paper stock exploded through the walls and ceiling of the mill, parts of which landed two hundred feet away. Another piece of the boiler burst backward through four walls and into the engine room. Within minutes, the smoldering ruins caught fire and spread rapidly to the five-story "rag" building. Since the wind was calm, the fire did not spread to the lumberyards, and the firemen were able to get the flames under control in about three hours. Remarkably, only nine of the twenty-five men on site were hurt, and none was killed. However, the building, machinery, raw stock and finished paper were all a total loss, and the corporation suffered $100,000 in damages with only $36,000 covered by insurance. Fortunately for its employees, the mill was rebuilt and back in operation by the following June.

There are several instances in which whole industrial settlements have succumbed to fire. Although the woodenware factories of Edinburgh had periodically suffered damage from fire throughout the 1870s and '80s, a devastating string of fires swept through the communities of Beecher's Hollow, Batchellerville and Tennantville in the late 1880s that greatly discouraged the mill owners. Most were not rebuilt, and about one-third of the residents moved elsewhere to find steadier work.

Milton Center was another casualty of fire. During the mid-1800s, it was one of the most prosperous hamlets in town north of Ballston Spa, containing two hotels, sawmills, gristmills, a wheel factory and one of the county's largest tanneries. In 1881, the tannery was completely destroyed, and the owner, Samuel Haight, decided to move the operation south to Ballston near the Delaware and Hudson Railroad mainline. Some years later, both hotels burned down, as well as several houses. By 1890, only the church and a few residences remained.

Two poorly timed and devastating fires within one year of each other caused a once thriving community to become depopulated at the turn of the century. Isaiah Blood's scythe and axe factories had successfully operated north of Ballston Spa since 1824. In 1892, the works had been purchased by a consolidation of hard edge toolmakers called the American Axe and Tool Company, which continued to run it much as it had been prior to the takeover. A disastrous fire devastated the scythe shop in March 1900. The cause of the fire was from an exploding lantern held by Thomas Jones, a foreman who was inspecting the oil pump. The explosion ignited the oil that was used as fuel for the welding department, and the flames quickly spread throughout the shop. When the fire reached the paint shop, the workers knew all was lost since there were large quantities of highly flammable turpentine and paint stored inside. When the fire had burned through the walls of the building, wind carried burning embers to the north, and soon the boiler house, grind shop, forge shop, blacksmith shop, coal shed and several storehouses were all in flames. Six horses that were stabled in one of the barns were saved, but all of the stock and most of the tools and machinery were completely destroyed. Although the shop possessed fire hose and other firefighting apparatus, the building that contained them was among the first to catch fire, leaving many of the workers to stand by helplessly and watch

A young boy surveys the ruins of the Ballston Scythe Factory in 1900. *Courtesy of the Chris Morley Collection.*

their place of employment burn uncontrollably. The flames were so high that many people from Ballston Spa could see them and walked two miles to witness the destruction, and telephone calls from as far away as Saratoga came in to ask what was burning. The loss was estimated at $70,000.

The tragedy of the scythe shop was followed only a year later in June 1901 with another destructive fire, this one at the axe works. The fire started at the west end of the main building and spread very rapidly throughout the complex. Night watchman James Barrett gave the alarm at 5:30 p.m., just after the workmen had left for the day. The men residing in the neighborhood at once responded and put the pump to work, with a line of hose strung from the bridge. Despite their quick response, the fire quickly engulfed the entire building. Efforts were made to save the blacksmith shop and engine room, which were separated from the forge shop by a narrow alley, to no avail.

This second fire was a serious blow to the hamlet of Bloodville and to the town of Milton as a whole. The scythe factory was still lying in ruins from the fire that had destroyed it the year before. Since the American Axe and Tool Company was gradually concentrating operations at its Pennsylvania plant, the second fire was the final nail in the coffin for the Bloodville factories. Although several attempts were made to locate other industries at the site over the next two decades, none lasted very long or brought anything like the prosperity that the axe and scythe works did. Bloodville soon became depopulated, and only later in the century would new housing developments replace the hundreds of people who once made the hamlet their home.

The village of Stillwater seems to have been especially unfortunate over the course of its history. The first major industrial-related fire occurred in 1816 when the Schuyler Mills burned to the ground. It was one of the biggest employers in the village, and its destruction led to several ordinances that banned certain types of celebrations and the practice of carrying live coals from one dwelling to another. Sporadic fires were recorded in the following half century, but in the 1870s a series of devastating conflagrations visited the village as if it had been given a curse. Stores, houses, hotels, a covered bridge and the Pemble Mills were all destroyed. A Button hand engine with hose was purchased from the Button Fire Engine Works of Waterford for $1,200 in an attempt to slow down the destruction.

The presence of the Button pumper was not enough to save the rebuilt Pemble Mills in 1884. The strawboard mill was discovered to be on fire on the morning of July 3, and firemen were soon on the scene. On the west side of the mill was Mosher, Judd and Company's paper mill, on the east side was R. and H. Newland's knitting mill, while just in front was the box factory of James Newland. Just as the D.E. LaDow steamer from Mechanicville arrived on the scene, another fire was discovered at a residence to the north of the Newland mill that threatened to ignite a large collection of wooden buildings clustered nearby. It was believed that a large portion of the village would have been razed if not for the sustained effort of the firemen from the two companies and many citizens. As it was, the Pemble Mill suffered $25,000 in damages and lost forty-six tons of paper.

The Pemble Mill fire led to the purchase of the village's first steam fire engine in 1886. However, the vehicle lost its own home in 1896 when the firehouse burned to the ground. One of the firemen was forced to crawl into the building on all fours and push the steamer onto the street. At the Newland and Denison mills, an automatic sprinkler system was installed in the summer of 1904. However, fire broke out before the system could be hooked up to the water main, and the mill and two adjacent knitting mills burned to the ground. In 1915, the Pemble Mill was again struck by fire and was all but

Fire struck the factories of Stillwater many times, such as the Saratoga Knitting Mill and the adjacent Newland knitting mill. *Courtesy of the Stillwater Town Historian.*

destroyed when the valve on the fire engine failed to work properly. It seems that time and again industries were established and hope for a prosperous future reappeared, only to be destroyed by the plague of fire.

Acts of arson have been recorded from the county's earliest days, beginning with the Tory and Indian raids of the late 1700s and lasting to the present day. The Pioneer Paper Company mill in West Milton was involved in one of the more interesting sagas in the county, which led to its complete destruction. The long-lived mill was probably first used as a paper mill around 1850, when it was sold to Chauncey Kilmer, owner of a paper mill in Rock City Falls, who then sold it to Coe Buchanan. The latter added several partners to the operation, such as Elisha Comstock, William Wilson and Solomon Parks. An article in the *Ballston Democrat* dated February 13, 1866, reported that the mill was following Kilmer's Stone Mill in using straw to manufacture newsprint for the *New York Herald World*, *Albany Express* and the *Troy Whig*. At the time, Marcus Comstock was the foreman, overseeing the production of 4,500 pounds of paper per day. The article quoted Comstock as saying,

"We are happy to state that the late unhappy circumstances which entailed much loss to the company have been arranged, and that now, as before, we are the 'Pioneers' in the paper manufacturing industry of this town." No mention was made of what those unhappy circumstances could have been, whether economic or worker related. One source alleges that the success of the mill brought dissension among the owners, with Buchanan fighting for complete control. Elisha Comstock brought suit against the other partners and eventually prevailed. He obtained a loan from First National Bank and began operating it with a new partner. Buchanan then went back to court and got a reversal of the decree and resumed management. However, losses had piled up during the litigation, and the loan could not be repaid.

In 1869, First National Bank ordered the collection of a judgment that it held against the Pioneer Paper Company. Deputy Sheriff Electus Dye took possession of the mill and prepared to spend the night there. He dismissed the regular night watchman and agreed to maintain a watch along with Elisha Comstock. About 4:00 a.m., he was startled by a bright light and found that the straw shed was on fire. He ran to the other end of the mill and opened the door to the finishing room to find it ablaze as well. Within minutes, the entire mill was engulfed in flames and almost completely destroyed, as well as a large amount of finished paper waiting to be shipped. Total damages amounted to $60,000, with insurance covering only one-third of that. To make matters worse, the insurance company alleged that there was improper security at the time and therefore refused to make a payment. Investigators immediately concluded that an arsonist set several fires throughout the mill. The burned-out buildings lay idle for several years but eventually became part of George West's paper bag empire and were completely rebuilt.

Waterford's flour industry all but ended with the destruction of the Shattemuck Mill in July 1869. The fire started in the "barrel house," which contained six thousand empty flour barrels. The nearby stable was full of hay and ignited with such force that the heat rapidly spread to the main mill building. The five-story brick structure was at times completely hidden by a wall of flames. Steamers from the McIntyre Hose Company of Cohoes, the King and Knickerbocker Engine Company of Waterford and the Tiuney Hose Company of West Troy worked for hours to get the flames under control, but the mill was completely destroyed. A.D. Shepard and Company reported a loss of buildings worth $50,000 and contents of $30,000, consisting of 10,000 bushels of feed, 8,000 bushels of wheat and 1,500 barrels of flour. Fifty men had been employed turning out 400 barrels of flour per day. The flour mill of House and Fulton was destroyed the previous year and not rebuilt, so the only flour and gristmills left in the village were those of J.B. Enos and Company and T.M. Vial and Sons. In fact, the knitting industry almost literally took the place of the flour industry in Waterford—the Massasoit Knitting Mills were built on the site of the Shattemuck Mill, the Diamond Knitting Mill was built on the former House and Fulton Flour Mill and the Waterford Knitting Company factory was refitted from the Rock Island Flouring Mills.

On May 30, 1884, the *Saratogian* exclaimed "Waterford's Calamity—Five Men Blown into Eternity" in a headline describing the destruction of the Moore and Wilson Strawboard Factory. At about 10:30 p.m., a terrific explosion startled residents in the entire western and northern parts of the village. Rotary bleachers made of heavy iron were used in the process of bleaching wood pulp at the mill. One of the bleachers, weighing eight tons, exploded with such force that it was propelled into the air and fell 400 feet away in a wheat field. The company's storehouse, located 150 feet away, was also destroyed, while the foundries of Holroyd and Company and J.M. King and Company were heavily damaged. Of the nine men who were working in the building that night, five were killed, including one who was found on the roof of a nearby building. The 50- by 125-foot factory building was a complete loss.

The dozens of knitting mills situated around Waterford also did not go unscathed. One of the largest fires occurred in July 1903 at John Pynes's Eureka Knitting Mill. The fire was believed to have started in the cotton shed and spread quickly through the wooden building additions and into the brick production building. The village had several fire companies, but two of them, the Knickerbocker Steamer Company and the Ford Hose Company, were attending the Tri-County Firemen's Convention in Glens Falls at the time. It also didn't help that there was a delay in sounding the alarm, so that by the time the Kavanaugh Hook and Ladder Company arrived, the fire had been burning for twenty minutes and engulfed the south half of the mill. The heat was so intense that the shingled roofs of surrounding houses kept igniting. Pynes estimated his losses at $250,000 with insurance for $209,500. Twenty newly installed machines were destroyed, and four hundred men were thrown out of work.

Although fire caused considerable damage to dozens of factories, it was not much of a factor in the decline of manufacturing in Saratoga County. There were many instances when businesses burned to the ground, causing tens of thousands of dollars in uninsured losses, but were rebuilt again on a larger scale. Only when a business was facing severe competition or suffering from the effects of other outside forces did mill owners decide to shut down for good.

THE TRUST MOVEMENT

A rather abrupt movement took place at the turn of the century to consolidate many types of industries into national corporations or trusts. Credit for the concept of a "trust" lies with John D. Rockefeller, owner of Standard Oil. During the late 1800s, Rockefeller began buying all of his competitors in the oil industry until his company refined 90 percent of the nation's oil. In danger of violating the Sherman Antitrust Act, which gave government the power to investigate and break up monopolies, Rockefeller and his lawyers created a new entity called the Standard Oil Trust. The corporation acted as a holding company, controlling the interests of other companies through stock ownership, which in turn controlled other companies. The public was understandably skeptical of the new type of corporation, but soon other industries followed suit and set up their own trusts to shut out competition. Dozens of trusts were created between 1895 and 1905, including the copper, steel, sugar, tobacco, leather, tool and paper trusts.

The most successful industries of Saratoga County and the people who worked for them were directly affected by this movement as local factories were absorbed or purchased by investors who almost always lived outside of the state. Sometimes this worked out well for the local economy, as the largest of these factories became one of the trusts' "flagship" mills, such as the paper mills in Corinth and Mechanicville and the tannery in Ballston Spa. However, there were other examples of local plants shutting down so that operations could be consolidated elsewhere.

One of the largest corporations formed at this time was the International Paper Company (IP) in 1898, known as the "Newspaper Trust." It was

The mammoth paper mill complex in Corinth was purchased by the International Paper Company in 1898. *Courtesy of the Corinth Town Historian.*

incorporated in New York City with $45 million in subscribed stock and $10 million in bonds. Three local paper mills were absorbed by IP: the Hudson River Pulp and Paper Company of Corinth, the Glens Falls Paper Mill Company of Glens Falls and the Lake George Paper Company of Ticonderoga. Fifteen other paper mills from around the northeastern United States also became part of the new company, and others followed suit in the years that followed, making IP the single largest paper manufacturer in the world. The Hudson River Pulp and Paper Company benefited from the change in ownership, as IP invested large sums of money to modernize the mill. It only closed down in 2002 when demand for its coated papers declined and the costs of maintaining the huge factory became too high.

Another beneficiary of the consolidation movement was the Duncan paper mill in Mechanicville. The West Virginia Pulp and Paper Company (Westvaco) was founded in 1899 by William Luke, who began a series of paper mill acquisitions, including the Duncan Company mill. In some ways, the community was negatively impacted by the change, such as when all of the former Duncan managers were dismissed and wages were periodically slashed. However, millions of dollars were invested into the plant by Westvaco over the following seven decades, and more than one thousand people enjoyed fairly steady employment.

The American Axe and Tool Company was established in 1889 in Oakland, Maine. That year, the company persuaded fifteen other companies to join it as part of the "Axe Trust," including the Ballston Scythe, Axe and Tool Works, operated by Isaiah Blood's son-in-law Henry Knickerbocker. During the 1890s, the trust reportedly controlled 90 percent of the country's production of axes and employed ten thousand people. After two separate fires destroyed the scythe factory and the axe works within a year of each other in 1900, the decision was made to concentrate operations at a new Glassport, Pennsylvania plant. The trust was not a strong one, however, and within ten years it began to break up. The American Axe and Tool Company finally dissolved in 1927 and was absorbed by the Kelly Axe and Tool Company. The American Fork and Hoe Company then acquired Kelly Axe, and in 1949 it was renamed True Temper Corporation. That company retained all of the rights to producing the former American Axe and Tool brands, including "I Blood."

The "Upper Leather Trust" consolidated tanneries that primarily produced leather for shoes and boots. The American Hide and Leather Company was the largest tannery conglomerate in the world when it took over the Hall, Haight and Company tannery of Ballston Spa in 1899. In addition to Hall, Haight and Company, American Hide owned twenty-one tanneries—thirteen in Boston, four in New York and four in other states—representing 75 percent of the upper leather business in the country. The trust was a financial success for over a decade until leather began to be replaced with substitutes like imitation leather or cloth. Other large tanneries were built around the country over the next few decades so that the Ballston plant could no longer be called the nation's largest, but it remained profitable for many years. High demand for leather after World War II created huge profits, and in 1947 the company paid its first dividends. However, the trust could not keep up with the constant changes in both technology and the market, so it was disbanded in 1956.

Another of the large consolidations involved the profitable paper bag industry. In 1869, the Union Paper Bag Machine Company was formed to buy and fight patents related to manufacturing paper bags. George West of Ballston Spa was one of the founders, and since there were only eight bag makers in the country at the time, this trust had a monopoly on the trade for many years. When combined, the member companies produced 371 million paper bags annually. At the time, the bag industry was dominated by two markets, one for standard paper bags and the other for "self-opening" paper bags. The rapid growth of the International Paper Company threatened

The Union Bag and Paper Company acquired all of George West's paper mills in 1899, including the extensive Union mill property in Ballston Spa. *Courtesy of the Chris Morley Collection.*

smaller companies with extinction, and its attempt at entering the paper bag industry alarmed those mill owners who produced them. To answer this threat, the two types of bag sectors were consolidated in 1899 when the Union Paper Bag Machine Company was dissolved and became the Union Bag and Paper Company. The new corporation had $27 million in capital and sold 4 billion bags annually—90 percent of the total United States market. In addition to the bag factories, the company owned extensive tracts of woodlands as well as water powers on the Hudson River at Sandy Hill. All of the mills of George West were purchased for $1.5 million.

Although Union Bag was not nearly as large as International Paper, it was by far the largest paper bag company in the world. The multitude of patents it acquired during the consolidation gave it some measure of defense against its competition. The effect on Ballston Spa, while promising at first, proved to be disastrous. Soon after George West passed away in 1901, Union Bag sold the Middle Grove mills, citing increased transportation costs to

Machines were installed in the Chase Bag Company's Fenimore mill to print its own bags. *Courtesy of the Eric Unkauf Collection.*

bring finished goods twelve miles by trolley railroad to the Delaware and Hudson Railroad interchange. The mills of Rock City Falls and West Milton followed suit. In 1915, the directors announced that they were putting all of the Ballston mills up for sale and moving its local bag production to Sandy Hill. In 1956, Union Bag merged with the Camp Manufacturing Company, which was later acquired by International Paper and passed out of existence.

In 1891, New Jersey businessman David Walton formed a folding box trust called the National Folding Box Paper Company. Two other partners named Thomas Cornell and Edward Shelton had patented a method for creating "knock-down" cardboard boxes that could be shipped flat in large bundles and then easily folded into a sturdy box. The method was highly successful, and in fact was copied by other manufacturers. The company decided to defend this patent and others it held in a series of lawsuits, instigating thirty cases from 1892 to 1902. These suits enabled the firm to control the small but lucrative market for "knock-down" boxes, primarily used for selling women's dresses. At its formation, the trust purchased the former McLean paper mill in Milton from Donaldson and Geer and converted the machinery from manufacturing straw print to folding cardboard boxes. The mill became a fixture in Factory Village, operating for thirty years almost without incident and providing much-needed jobs to local residents. The mill only shut down after a long workers' strike in 1919.

In 1902, the Thomson Pulp and Paper Company of Schuylerville became one of dozens assumed by the United Box Board and Paper Company. The New Jersey–based company had a capital of $29 million and, after the consolidation, boasted an output of 1,359 tons of boxboard, ground wood, sulphite and soda pulp annually. It used five one-thousand-pound engines and one 120-inch twelve-cylinder engine to produce 25 tons of jute fiber board per day. After a series of disputes between the stockholders and management, the mills reverted back to the original owners, but in 1912 there emerged a new trust called the United Paperboard Company. Six ground wood mills were reacquired, including the one in Schuylerville, and eleven paper board mills.

The long-lived Button Fire Engine Works was one of the area's most specialized industries. There is a record of dozens of companies formed to manufacture fire engines around the country that only produced one to ten engines. Lysander Button's many inventions allowed the company to prosper as technology rapidly changed and many other manufacturers went bankrupt. However, after spending almost half a century in business, Button decided to retire in 1882 and sell the entire works, including the rights to his many patents, to Holroyd and Company of Waterford. A fire engine trust was formed in 1891, called the American Fire Engine Company, which purchased the Button Fire Engine Company, the Clapp and Jones Manufacturing Company of Hudson, the Silsby Manufacturing Company and the Ahrens Manufacturing Company. The industry was further consolidated in 1900 when a group of New York investors formed the International Fire Engine Company. Unfortunately, the manufacture of fire engines in Waterford was discontinued shortly after the business left the hands of James Holroyd, and the factory building was sold to Novelty Knitting in 1908.

It is difficult to tell if the short-lived but far-reaching trust movement had a positive or negative impact on Saratoga County. It appears that in some cases, communities did benefit from the consolidation of a local mill with others around the country. The paper mill in Corinth was continually updated and enlarged during its life as part of International Paper, as was the Mechanicville paper mill of West Virginia Pulp and Paper. Both mills are now closed, but it is possible that they would have closed down sooner if they did not enjoy the economies of scale received by being part of a larger company. Conversely, when the Union Bag and Paper Company purchased the mills of George West, it began the process of shutting them down, while the fire engine industry in Waterford died almost immediately after merging with a trust. The trend of moving industry to western or southern states would eventually claim the vast majority of mills in the county, whether they were part of a trust or not, so the effects of the trust movement will have to be left to debate.

Remnants of a Bygone Era

Author James Kunstler wrote a controversial article for the *New York Times* in 1990 titled "Schuylerville Stands Still," in which he detailed the loss of industry and its effects on the village. Although his observations may have led some readers to believe that Schuylerville was on the edge of societal ruin, his conclusion that small, formerly industrial communities must reinvent themselves in order to sustain a healthy and happy population has merit.

Some communities in Saratoga County have been more successful than others. Mechanicville, Corinth and Schuylerville are still dealing with the loss of their heavy industry today, especially because some of the biggest mills only closed down within the last few decades. Other communities have been able to slowly recover due to the construction of the Adirondack Northway, which gave residents the ability to commute to Capital District cities for work.

Areas of the county that were never industrialized, such as Clifton Park, Halfmoon and Malta, have rapidly become major population centers due to their proximity to Albany, Schenectady and Troy. Saratoga Springs never depended upon manufacturing to sustain its economy and so has been able to maintain its wealth by promoting tourism that centers on its racetrack.

By comparison, the town of Stillwater experienced a steep decline in population when the West Virginia Pulp and Paper Company mill in nearby Mechanicville closed in 1971. At the time of its closure, the population was 6,247, which dropped to 5,500 ten years later. In the last few years, it has rebounded as more and more people find the town to be a desirable residential community.

Today, manufacturing makes up less than 10 percent of all economic activity. There are approximately 9,500 manufacturing jobs spread out around the county, as opposed to over 15,000 retail-wholesale jobs, 21,000 government jobs and 25,000 educational-healthcare jobs. Almost half of the 108,000 total number of workers in the county are employed in management and professional occupations, while another 40 percent are in the service and sales occupations.

Ultimately, the industrial base of Saratoga County suffered the same fate as other northeastern manufacturing centers. Beginning in the late 1800s, new concentrations of raw materials were found in the western part of the country, while southern states offered a convenient supply of cotton as well as inexpensive labor. The Capital District as a whole saw its once vaunted industries slowly relocate or shut down, including its iron and steel foundries, stove factories and clothing mills. Those that survived the Great Depression, such as General Electric and American Locomotive, scaled back or shut down later in the century.

Most of Saratoga County's industries shut down even sooner. Many of its mills were built in the age of water power and were only able to prosper as long as other mills in the country were also powered by water. Large, urban-centered factories powered by electricity were more efficient by an order of magnitude, driving down the prices of most manufactured goods. Tariffs that once protected American goods from cheap foreign-made products were gradually lifted, further lowering the prices of paper, steel, clothing, tools and other items manufactured in the Capital District. The simultaneous rise of labor unions, which sometimes dramatically increased the cost of doing business, also had its toll. The small mills on the Hudson, Mohawk and Kayaderosseras Rivers began losing money and were forced to close.

The first to go were the smaller factories that were unable to compete on a national or international scale. The increased use of railroads in the late 1800s meant that most businesses could sell goods inexpensively anywhere in the country. Only those that merged with other manufacturers or specialized in some way were able to survive. Local sawmills and gristmills, which traditionally served adjoining communities, were replaced by those that served entire regions. The small woodenware factories also disappeared, unable to compete with modern factories.

Most of the paper mills were the next to close. The comparatively small mills of Hadley, Ballston Spa, Stillwater and Schuylerville were unable to compete with International Paper and other paper trusts. With a few exceptions, only the two mills that were themselves part of the trust movement, located in Corinth and Mechanicville, continued to remain competitive.

The clothing industry was the last great industry in the county to disappear, lasting well into the 1900s in Mechanicville, Ballston Spa and Waterford. Only the availability of cheap labor in the South and inexpensive clothing from foreign countries caused the factories to be relocated or shut down. Ironically, the last clothing factory in Ballston Spa was forced to close when it was unable to find enough people to work for the wages it was able to offer.

Despite the loss of nearly all of its industry during the twentieth century, some manufacturers were able to reinvent themselves or find a way to specialize and remain competitive on the world stage. For example, the huge paper mills that dominated the landscapes of multiple communities are almost gone, but a few remain competitive in the niche that they have carved out for themselves. Sweden-based SCA Tissue purchased and renovated the South Glens Falls mill buildings in 2001, as well as a plant in Greenwich. The mill employs about 260 hourly workers represented by the United Steel

The office building of the Glens Falls Paper Mill Company is now used by SCA Tissue. *Courtesy of the author.*

Workers Union and 60 salaried workers. SCA uses all recycled paper to manufacture tissue paper, amounting to eighty thousand tons of product per year. The corporate entity owns six other manufacturing facilities and is the third largest manufacturer of "Away From Home" tissue products. In 2009, the company announced that it was investing $20 million into its South Glens Falls mill to construct a new fifty-thousand-square-foot building for its new Torkmatic Industrial Rewinder machine and other upgrades.

The other large paper mill in the county is located in Waterford. The history of the Mohawk Paper Mills began in 1872, when several businessmen established a small paper mill at the old axe handle factory on King's Canal. At first it was called the Mohawk and Hudson Paper Company, but Frank Gilbert, one of the original partners, purchased the mill outright and changed the name to the Frank Gilbert Paper Company. Within a few years, the factory was producing three tons of printing paper per day and employing forty people. In 1917, the company expanded by adding a second paper mill in Cohoes near the junction of the Champlain and Erie Canals. This mill manufactured a variety of paper, including mimeo, bond, ground wood and wallpaper.

The Great Depression caused the paper mill to declare bankruptcy in 1930. Waterford lawyer George O'Conner was appointed receiver and reorganized the company under the name Mohawk Paper Mills, Inc. Since O'Conner knew little about the paper industry, he invited the International Brotherhood of Papermakers to organize a union of two hundred workers. Maintaining a good relationship with the new union was instrumental in remaining viable throughout the Depression. The advent of World War II finally provided an opportunity for the company to turn a profit. Most of the product line was made up of box cover and box lining papers. Soon after the war, the mill began producing such paper as Monopoly money and box coverings for Old Spice. The company also became well known for its Halfmoon Vellum, Poseidon, Cortlea, Artemis and Superfine grade of papers that were used in commercial printing. One remarkable feat performed by the company was increasing its annual output from nine hundred tons annually in 1880 to ninety thousand tons annually one hundred years later using the same three machines.

In 1969, Riegel Paper Company purchased the mill, which was taken over by Federal Paperboard. However, three years later, Thomas O'Conner bought the company back, and over the next two decades the O'Conner family infused the plant with millions of dollars in investments. The company continues to adapt itself to changing technologies to the present day. In 1999, it was granted a patent for its Inxwell technology that provides uncoated paper with better print quality than previously possible.

A third paper mill operates on a smaller scale in the town of Milton. In 1926, Frank Cottrell purchased the Kilmer Stone Mill in Rock City Falls after a series of owners tried to make it viable over the course of twenty years. Of the two dozen mills that once operated along the Kayaderosseras Creek, the Cottrell Paper Company is the only mill still in operation today. It produces insulation for electrical transformers and "fractional horsepower" electric motors (less than one horsepower). It is evident that the Cottrells were "forward thinking" in several ways. They concentrated on a little-known but necessary product line for the rapidly rising usage of electric technology. From the start, they recognized the importance of keeping the Kayaderosseras Creek as clean as possible to avoid lawsuits from those living "down the river" in Ballston Spa and elsewhere. They were also the first to fully realize that truck transportation would very soon replace the railroads and began to use the new highways immediately for distributing its products, despite the fact that the electric railroad was still running through its property. The secret to the mill's remarkable longevity lies in the process it uses, which turns out a high-quality paper. The raw materials are 90 percent recycled, provided by jeans manufacturers. The denim is cut up and "cooked" in caustic soda to break up the cotton and then put into blenders to break it down further into pulp. The Kayaderosseras provides all of the vast amounts of water needed for the operation, which means historians are able to see the dam and pond as it looked in the late 1800s. The huge wooden vats that purified the water have been replaced with a steel filtration system, and other systems have been updated, but the process for making the paper remains largely unchanged.

A remnant of Saratoga Springs's resort and mineral spring era can be found just south of the city on Geyser Road. The Saratoga Spring Water Company was founded in 1872 to bottle and sell mineral water from the Sweet Water Spring. The original spring was hand-drilled through 30 feet of sand and 150 feet of rock, and it still flows today with natural carbonation. Its products include sparkling and non-sparkling spring water, Saratoga Essence for upscale establishments and the fruit-flavored Saratoga Splash, all of which are sold around the country.

During the 1800s, a variety of factories provided work to over 750 people each, some even employing over 1,000. Today the only employers of 1,000 or more are entities such as Saratoga County government, the Schenendehowa Central School District and State Farm Insurance. The largest manufacturers are SCA Tissue and Mohawk Paper Mills discussed above. Ball Metal Container (based in Colorado and occupying a 290,000-square-foot plant in Saratoga Springs) and Espey Manufacturing each employ several hundred people. Espey was founded in Saratoga Springs and designs, develops and manufactures magnetic components and power conversion electronics for military applications. Cascades Tissue Group, which produces tissue paper, employs just over 200 people between its Waterford and Mechanicville mills. All other manufacturers employ fewer than 200 workers.

The hand of time has not been kind to the industrial history of Saratoga County. Many of the factories and mills that once drove the economy have disappeared without a trace; others leave only small clues to their previous existence. Some survive to this day either converted to other uses or vacant, while a scant few remain relatively unchanged over the past century and in use for industrial pursuits.

Nearly all of the earliest mills disappeared many years ago. Sawmills, gristmills, tanneries and shops were all made of wood. When their useful lives ended, they could not be easily converted to other uses and were torn down. It is sometimes difficult for the trained eye to spot exactly where some of these mills were situated, even if it is known that they existed next to a particular stream or road. A few locations in Galway, Charlton, Milton, Saratoga, Clifton Park and other towns offer clues to the existence of small mills in the form of old stone dams and artificial sluiceways, but little else. For example, researchers in Charlton uncovered the remnants of a small milldam on the Alplaus Kill near Eastern Avenue. Later they found a raceway chiseled in the stone and pieces of metal that indicated the presence of an undershot water wheel. These remains turned out to be what was left of the Truax sawmill and gristmill, erected in the late 1700s. Dozens of remains such as this one can be found throughout the county, as many of the early mills were situated on small streams away from present-day paved roads.

However, even some of the largest mills in the county have left behind little evidence. Most of the woodenware factories along the Sacandaga River were submerged with the creation of the reservoir in 1930. Most of the paper mills along the Kayaderosseras Creek in Milton and Greenfield have vanished despite the lack of any further development on those sites. The entire Greenfield village of Mount Pleasant, constructed to serve the Granger Glass Factory, is no more than a clearing north of Lake Desolation.

A few of the largest mills that acted as the main engine for the local economy stand vacant as of this writing. The mammoth International Paper complex in Corinth, once known as the Hudson River Pulp and

Paper Company, is perhaps the most recognized of these. The Victory Mill outside of Schuylerville was abandoned years ago by its creator. Today, the complex towers above the village as an almost cruel reminder of more prosperous times. The brick portion of the Anthony Spice Mill in Halfmoon still stands in excellent condition off Farm-to-Market Road, surrounded by new housing developments. The giant Tufflite Plastic building in Ballston Spa, originally constructed to house the Ballston Refrigerating Company, is partially used as an antique store, while the rear part of the building on Bath Street has begun to crumble.

Nearly all of the various mills along the King's and Champlain Canals in Waterford are gone; most of the mills were torn down years ago for redevelopment or for safety concerns. However, there is ample evidence of the village's industrial past. Remnants of the canals can be seen throughout the village; several of the mansions and town houses of the industrialists have been preserved; and several of the brick mills have survived and been put to other uses. A five-story brick building on Mohawk Street is now used by Ursula of Switzerland but formerly housed the Waterford Knitting Company, the Ormsby-Morris Company and the Laughlin Textile Mill. The Mohawk Paper Mill operates nearby as one of only four paper mills that have survived from the dozens that once existed.

Other former mills have been transformed into spaces appropriate for modern use. The bag factory of George West's paper bag empire was turned into apartments, while his Union Mill next door is now a retail business and restaurant complex. The Cluett, Peabody and Company bleachery on Peebles Island is now part of a state park. The G.F. Harvey Company in Saratoga Springs is a professional building. Down the road on Route 50, the former Baker and Shevlin Foundry was acquired by Espey Manufacturing.

The southern tier of the county, particularly Halfmoon and Clifton Park, has experienced incredible growth over the last few decades. The great population shift from the cities to the suburbs that took place after World War II has continued as more people choose to leave Albany, Schenectady and Troy for the relative peacefulness of Saratoga. The growth has been so dramatic that it has been challenging for many local governments to adapt to the influx of new citizens. As land in the southern portion of the county became increasingly more expensive, other towns such as Milton, Malta and Wilton have experienced dramatic growth as well. Saratoga County is now the second most populous county in the Capital District area, with over 210,000 residents, compared to 295,000 in Albany, 152,000 in Rensselaer and 146,000 in Schenectady. All other counties in the region have fewer than 75,000 people.

As this book was being written, Global Foundries was beginning work on a new microchip fabrication plant in the Luther Forest Technology Campus, located in the town of Malta. The 1.3-million-square-foot facility is expected to usher in a new age of manufacturing in Saratoga County as other companies move in to support the plant and its workers. A "Tech Valley" similar to those in California and Texas may one day give employment to thousands of people, harking back to a time when the paper and knitting mills in Saratoga County gave employment to thousands of hardworking citizens.

Bibliography

Alling, Harold. *New York State Museum Bulletin: The Adirondack Graphite Deposits*. N.p.: University of the State of New York, 1918.

American City. Published monthly by the Civic Press, New York, 1920.

Bastin, Edson. *Feldspar Deposits of the United States*. Washington, D.C.: Department of the Interior, Washington Government Printing Office, 1910.

Bolles, Albert. *Industrial History of the United States: A Complete Survey of American Industries*. N.p.: Henry Bill Publishing Company, 1878.

Bulkeley, John. *Leading Men and Leading Pursuits of Ballston and Vicinity*. Ballston Spa, NY: printed by W.S. Waterbury, 1874.

Clapperton, George. *Practical Paper-making*. London: Rosby Lockwood and Son, 1907.

Clifton Park–Halfmoon, A Community Study. Published by the League of Women Voters of Shenendehowa, 1973.

Cochrane, Charles. *Modern Industrial Progress*. Philadelphia: J.B. Lippincott Company, 1904.

Corbett, Theodore. *The Making of American Resorts: Saratoga Springs, Ballston Spa, Lake George*. New Brunswick, NJ: Rutgers University Press, 2001.

Davenport, Bishop. *A History and New Gazetteer of North America and the West Indies*. N.p.: J.W. Benedict and Company, 1843.

DeGarmo, Todd. *Life in a Mill Town*. N.p.: published by the Brookside Museum, 1990.

Donaldson, Alfred. *A History of the Adirondacks, Volume II*. New York: Century Company, 1921.

Dunn, Violet. *Saratoga County Heritage*. Saratoga County, NY: 1974.

French, J.H. *Gazetteer of the State of New York*. Syracuse, NY: R. Pearsall Smith, 1860.

The Gristmill. Quarterly newsletter published by the Brookside Museum.

Grose, Edward. *Centennial History of Ballston Spa*. Ballston Spa, NY: Ballston Journal, 1907.

Hammersley, Sydney. *A History of Waterford, New York*. Waterford, NY: 1957.

Johnstone, Jan. *Saratoga County Communities: An Historic Perspective*. N.p.: Saratoga County Planning Board, 1980.

Larkin, F. Daniel. *Pioneer American Railroads: The Mohawk and Hudson and The Saratoga and Schenectady*. N.p.: Purple Mountain Press, 1995.

Lockwood's Directory of the Paper, Stationery and Allied Trades. New York: Lockwood Trade Journal Company, 1906 and 1919.

Mineral Resources of the United States. Washington, D.C.: Department of the Interior, United States Geological Survey, Washington Printing Office, 1908.

Munsell, Joel. *A Chronology of Paper and Paper-Making*. Albany, NY: 1870.

Newland, D.H. *New York State Museum: The Mining and Quarrying Industry of New York State*. N.p.: University of the State of New York, No. 426, 1908.

O'Shea, Amelia. *The Erie Canal Through Saratoga County*. Freehold, NY: 2007.

Report of the Sanitary Engineering Division. New York: New York State Department of Health, Twenty Ninth Annual Report, J.B. Lyon Company, 1909.

Saratoga Illustrated: The Visitor's Guide. New York: Taintor Brothers and Company, 1875.

Shaughnessy, Jim. *Delaware and Hudson: The Bridge Line to New England and Canada*. Syracuse, NY: Syracuse University Press, 1997.

State Board of Health, New York: Annual Reports 1896–1910. Albany, NY: published by J.B. Lyon Company, 1910.

Sylvester, Nathaniel Bartlett. *History of Saratoga County, New York (1609–1878)*. N.p.: Everts and Ensign, 1878.

Thomas, Lester St. John. *Timber, Tannery and Tourists: Lake Luzerne*. Lake Luzerne, NY: Committee on the Publication of Local History, 1979.

Vanderwerker, J.B. *Early Days in the Vicinity of Northumberland and Bacon Hill*. N.p.: 1938.

Weeks, David Chapin. *Saratoga County at Work 1865–1918*. N.p.: published by the Brookside Museum, 1979.

Weeks, Lyman Horace. *A History of Paper Manufacturing in the United States 1690–1916*. N.p.: Lockwood Trade Journal Company, 1916.

Index

D

E

F

G

H

I

K

L

M

V

W

About the Author

Timothy Starr was born in Danbury, Connecticut, and moved with his family to Hebron, New York, at the age of six. During his senior year in high school, he wrote his first book, a novel called *The Meatloaf Incident and Other Adventures* that is based on his experiences growing up in a rural community. He received a bachelor of science degree in accounting and a bachelor of arts degree in history from Castleton State College. He also received a master's degree in business administration from Chadwick University.

Soon after graduating college, he moved to the historic county of Saratoga. After noticing many ruins around the town of Milton, he began researching its history, particularly during the mid- to late 1800s. Starr wrote his first nonfiction book in 2007, titled *Lost Railroads of the Kaydeross Valley* (out of print). That project led to another one that detailed the industries served by the railroad, called *Lost Industries of the Kaydeross Valley*. When he discovered that several important inventions were developed to support these industries, yet another project unfolded that resulted in a Ballston Spa book of inventions and a Saratoga County book of inventions. His two biographies cover the lives of local industrialists George West and Isaiah Blood.

For most of his career, Mr. Starr has been involved in accounting and finance in the nonprofit industry. He has contributed chapters to several local history books and is a frequent contributor to the *Ballston Spa Life* "History Lesson" series. He also volunteers as the treasurer of the Saratoga County Historical Society (Brookside) Board of Trustees. He lives in Rock City Falls with his wife, Alison, and daughter, Morgan.

Other books by the author

The Ballston Terminal Railroad and Its Successors
Great Inventors of New York's Capital District
Invented in Ballston Spa
Invented in Saratoga County
Isaiah Blood: Scythe and Axe Maker of Ballston Spa, New York
Leading Industrial Pursuits of Ballston Spa, Glens Falls, Sandy Hill & Fort Edward
Lost Industries of the Kaydeross Valley: A History of Manufacturing in Ballston Spa, New York
The Paper Bag King: A Biography of George West